THE
SOCIOLOGICAL
SPIRIT

THE SOCIOLOGICAL SPIRIT

SECOND EDITION

EARL BABBIE

Chapman University

Wadsworth Publishing Company
Belmont, California
A Division of Wadsworth, Inc.

Sociology Editor: Serina Beauparlant
Editorial Assistant: Susan Shook
Production Editor: Julie Davis
Managing Designer: Andrew Ogus
Print Buyer: Karen Hunt
Permissions Editor: Jeanne Bosschart
Designer: Andrew Ogus
Copy Editor: Cheryl Ferguson
Compositor: Joan Olson, Wadsworth In-House Composition
Printer: Malloy Lithographing, Inc.

HM
51
. B164
1994

*This book is printed on
acid-free recycled paper.*

I(T)P™

International Thomson Publishing
The trademark ITP is used under license

Printed in the United States of America

3 4 5 6 7 8 9 10—98 97 96

Library of Congress Cataloging-in-Publication Data

Babbie, Earl R.
    The sociological spirit / by Earl Babbie
        p.   cm.
    Includes bibliographical references and index.
    ISBN 0-534-20202-0
    1. Sociology   2. Criticism (Philosophy)   I. Title
HM51.B164   1994   93-19545

To
Talcott Parsons
1902–1979
a master sociologist
and master teacher

CONTENTS

Chapter 3

Groups
46

Chapter 4

Organizations
62

Chapter 5

Institutions
79

Chapter 10

The Global Picture
164

PREFACE

Albert Einstein once characterized the advent of nuclear weapons by saying everything had changed except our ways of thinking about things. He was concerned that if we maintained our old ways of thinking, we would be in big trouble.

This little book grew out of the conviction that the need for new ways of thinking extends well beyond the nuclear arena. We need new ways of thinking if we are to handle the problems of world hunger, poverty, crime, prejudice, pollution, and all the other ills that plague modern life. There is an ancient Chinese proverb warning that if we continue in the same direction, we will end up where we are headed. I suggest that if we continue to think about society and its ills in the same way that we have been thinking about them, we will end up with more of the problems we currently face—except they will be worse.

My purpose in writing this book has been to examine some of the possibilities of sociology as a way to view, understand, and affect the world we live in. In Chapter 1, I present some examples of the ways we often go astray in dealing with social issues. In that chapter, I call sociology "an idea whose time has come," because I am convinced that sociology offers an important shift in thinking that is vitally needed in today's world. The sociological perspective can offer insights not readily available in other ways, and my purpose is to give you a sense of what the sociological perspective is.

In some sense, this book focuses on the paradoxical interplay of individuals and institutions. A key sociological insight is that society is more than a mere collection of individual human beings. Social institutions have a power that transcends and outlives the humans involved. The failure to recognize that can handicap you in dealing with the important problems and opportunities in the world and in your own life. However, recognizing the power of social institutions can be a handicap as well, reflected in the cry, "I'm just a little guy. What can I do?" My overarching purpose is to empower you to deal with the paradox of individuals and institutions.

This is not intended as a comprehensive introduction to sociology. There are a number of excellent texts available that provide exhaustive coverage of sociological concepts and theories and report on the most up-to-date research findings. Instead, this book can be used in an introductory sociology course in conjunction with a standard textbook or other readings to provide a global overview of sociology and set the tone for the course. Alternatively, this book might be used as a review at the end of the course, helping to integrate course themes. At the same time, I hope the book will be of value to those readers who are not taking any courses in sociology but would like to know a little about the sociological perspective. If you are such a person, I think you will find this little book a more accessible door to sociology than a full-blown textbook.

I began these comments by indicating that sociology can empower you to deal with the problems of the world. At the same time, I think you will find sociology directly relevant to your own personal life. Chapter 2, for example, deals directly with the question, "Who am I?" I think you will find yourself more deeply engrossed in that issue than ever before. As we'll see, the answer to the question is by no means obvious. Ultimately, the question is perhaps unanswerable, but we'll see that the search for answers can be very valuable. The remainder of the book examines the ways in which "who you are" is a function of the society in which you live—and we'll see how recognizing that can help you deal with society's constraints.

A number of people have contributed to this book, and I'd like to take this opportunity to thank some of them. First, a number of my sociological colleagues around the country were kind enough to spend time reading early drafts of the book and give me their comments. While they can't be held accountable for any of its persistent shortcomings, they certainly contributed to making it better. They are Connie Elsberg, Northern Virginia Community College; James Gallagher, University of Maine; James Long, Golden West College; Stephen Steele, Anne Arundel Community College.

By the same token, a number of people at Wadsworth Publishing Company played vital roles in the book's creation. I regret that it's probably a cliché to call a book like this a "team effort," but that's precisely what it is. Serina Beauparlant has been a strong and supportive editor, and I am grateful for her work on this revision. I hope she can see the impact she has had on it.

The Wadsworth production and design department has been responsible for transforming the pile of manuscript pages into a real book. My thanks go to Julie Davis, the production editor; Andrew Ogus, the managing designer; Karen Hunt, the print buyer; Alan Noyes and his crew in the production art department; and Cheryl Ferguson, the copy editor.

I have dedicated this book to Talcott Parsons. Although I could go on at length extolling his impact on sociology theory, my dedication is more personal than that. When I arrived at college as an undergraduate, I don't think I had even heard the term *sociology*. I learned what it meant in course after course with Parsons; I was so utterly fascinated by the new way of seeing and thinking that his lectures opened up for me that I just came back again and again. Even though we were never close personally, I have never lost any of the gratitude I first felt all those years ago. I hope I can share with you some of the excitement of sociology that Parsons shared with me.

CHAPTER I

AN IDEA WHOSE TIME HAS COME

There is a more pressing need for sociological insights today than at any time in history. In 1822, the French philosopher Auguste Comte first proclaimed the possibility of studying society scientifically. A century and a half later, sociology is an idea whose time has come, and not a minute too soon.

It is no secret that this generation faces several unprecedented dangers. No sooner has the Cold War seemingly ended than we have come to recognize the danger of localized wars among ethnic groups spilling over into wider, global conflict. This danger is made worse by the possibility that relatively small, impoverished nations could gain access to nuclear weapons, giving terrorists the opportunity to spark an international conflagration. The more we learn about the prospects of a "nuclear winter"—the likely result of the first truly nuclear war—the more evident it is that there would be no real survivors of such a large-scale nuclear exchange.

If, on the other hand, we escape the threat of nuclear extinction, there is a real possibility that we will overpopulate and pollute the planet beyond its carrying capacity. As a single indicator of this problem, some 13 to 18 million people die of starvation around the world every year, three-fourths of them children. Approximately one-fifth of all humans on the planet go to bed hungry every night.

Add to this such persistent problems as crime, inflation, unemployment, prejudice, totalitarianism, and national debts,

and you have sufficient grounds for understanding the ancient
Chinese curse: "May you live in interesting times."

These are unquestionably interesting times. But the picture
is not completely gloomy. These are also the times of great
achievements in space: humans landing on the moon, a
remote-controlled craft landing on Mars, and others pho-
tographing the more distant planets. These are the times when
human beings, working cooperatively around the globe, eradicated smallpox—a scourge throughout history. These are the
times of an awakening of awareness and commitment to ending world hunger. And the breakup of the former USSR, along
with its domination of Eastern Europe, is regarded by most as
a positive development.

If it were possible to make comparable lists of the positive
and negative aspects of life today, my hunch is that they
might be about equal in length. At any rate, both lists would
be long ones, indicating that you and I and our fellow human
beings face both trying challenges and promising opportunities ahead of us.

Assuming that we'd agree in favoring peace over war,
prosperity over hunger and poverty, and so forth, the question
you should be asking yourself is: what determines how things
turn out? Specifically, what would it take for peace to triumph
over the continuing specter of wars large and small? We'd all
like the answer to that question.

I suggest there is a prior question you should ask. That is:
where should we look for the answer to how peace can prevail
over war? Before asking what the answer is, we need to ask
where the answer is likely to be found. I suggest that up until
now we have tended to look for the answer to how peace can
prevail over war within the domain of military technology.
Most simply, we have tried to create weapons that would pre-
serve the peace.

When Hiram Maxim invented the first fully automatic
machine gun in 1884, he believed he had actually brought an
end to war. In his view, "Only a general who was a barbarian
would send his men to certain death against the concentrated

power of my new gun."[1] Instead of ending war, however, the Maxim gun simply made killing more efficient.

Some felt that the airplane would mark the end of war. No less an authority than Orville Wright said, "When my brother and I built and flew the first man-carrying flying machine, we thought we were introducing into the world an invention that would make further wars practically impossible."[2] Again, as the residents of Dresden, London, Pearl Harbor, Hiroshima, Nagasaki, and Baghdad were to learn, the airplane only made war more deadly.

More recently, many have felt that nuclear weapons were so horrible as to make war unthinkable. But although the combined U.S.–USSR nuclear arsenals at their peak equaled 6,000 times the power of all the explosives used in World War II, the two superpowers continued rushing to build more. Moreover, there have been countless wars around the globe since the invention of nuclear weapons.

The point of this discussion on war and peace is to suggest that what we need to know to establish peace around the world is not likely to arise from military technology. If such an answer is to be found at all, we must look elsewhere: in the study of why people relate to one another as they do—sometimes peacefully, sometimes hostilely. This, as we'll see, lies in the domain of sociology.

The Domain of Sociology

Sociology involves the study of human beings. More specifically, it is the study of interactions and relations *among* human beings. Whereas psychology is the study of what goes on

1. Hiram Maxim as quoted in Martin Hellman, *A New Way of Thinking*, Palo Alto, CA: Beyond War, 1985, p. 4.

2. Orville Wright as quoted in Martin Hellman, *A New Way of Thinking*, Palo Alto, CA: Beyond War, 1985, p. 4.

inside individuals, sociology addresses what goes on between them. Sociology addresses simple, face-to-face interactions such as conversations, dating behavior, and students asking a professor to delay the term paper deadline. Equally, sociology is the study of formal organizations, the functioning of whole societies, and even relations among societies.

Sociology is the study of how human beings live together—in both the good times and the bad. It is no more a matter of how we cooperate and get along than of how we compete and conflict. Both are fundamental aspects of our living together and, hence, of sociology.

You might find it useful to view sociology as the study of our *rules for living together*. Let's take a minute to look at that.

To begin, let's consider some of the things that individuals need or want out of life: food, shelter, companionship, security, satisfaction—the list could go on and on. My purpose in considering such a list is to have us see that the things you and I need or want out of life create endless possibilities for conflict and struggle. When food is scarce, for example, I can only satisfy my need at your expense. Even in the case of companionship—where both people get what they want—you and I may fight over a particular companion.

The upshot of all this is that human beings do not seem to be constructed in a way that ensures cooperation. Bees and ants, by contrast, just seem to be wired that way. As a consequence, human beings *create rules* to establish order in the face of chaos. Sometimes we agree on the rules voluntarily, and other times some people impose the rules on everyone else. In part, sociology is the study of how rules come into existence.

Sociology is also the study of how rules are *organized* and *perpetuated*. It would be worth taking a minute to reflect on the extent and complexity of the rules by which you and I live. There is a rule, for example, that Americans must pay taxes to the government. But it doesn't end there. The rule for paying taxes has been elaborated on by a great many more specific rules indicating how much, when, and to whom taxes are to be paid. In recent years, the index to the IRS tax code has run more than 1,000 pages long, which should give you some idea of the complexity of that set of rules. The much-touted tax simplification of 1986 was 1,855 pages long.

The rules governing our lives are not all legal ones. There are rules about shaking hands when you meet someone, rules about knives and forks at dinner, rules about how long to wear your hair, and rules about what to wear to class, to the symphony, and to mud-wrestling. There are rules of grammar, rules of good grooming, and rules of efficient computer programming.

Many of the rules we've been considering were here long before you and I showed up, and many will still be here after we've left. Moreover, I doubt that you have the experience of having taken part in creating any of the rules I've listed. Nobody asked you to vote on the rules of grammar, for example. But in a critical way, you *did* vote on those rules: you voted by obeying them.

Consider the rule about not going naked in public. Even though you don't recall being asked what you thought about that one, there was a public referendum on that issue this morning—and you voted in favor of clothes. So did I. If this seems silly, by the way, realize that there are other societies in which people voted to accept a different rule this morning.

Sometime today you are likely to be asked to vote on a set of rules about eating. Some of the possibilities are eating spaghetti with a knife, pouring soup on your dessert, and throwing your food against the wall. Let's see how you vote.

The persistence of our rules is largely a function of one generation teaching them to the next generation. We speak of **socialization**[*] as the process of learning the rules, and it becomes apparent that we are all socializing each other all the time through the use of positive and negative **sanctions**—rewards and punishments.

---

[*] Boldfaced terms are defined in the glossary.

All the rules we've been discussing are fundamentally *arbitrary*—that is, different rules would work just as well. Although Americans have a rule that cars must be driven on the right side of the road, other societies (e.g., England, Japan) manage equally well with people driving on the left side.

Once we've established a rule, however, we tend to add weight to it. We act as though it were better than the other possibilities, that it somehow represents an eternal and universal *truth*. Sociologists often use the term **reification** in reference to the pretense that things are real when they are not, and we often *reify* the rules of society. We make the right side of the road the *right* side for cars to drive on, and we think the British and Japanese strange for not knowing that.

The rules of society take their strongest hold when they are **internalized** by individuals—taken inside ourselves and made our own. Imagine this situation, which you may have actually experienced. It is three o'clock in the morning, and you are driving along a street leading out of a small town. There is no traffic on the street in any direction as you come to a red light at an intersection. You can see there are no cars coming for a mile in every direction. There is no one around. What do you do? There's a good chance that you will sit and wait for the light to change. If someone questioned you about it, you might say, "It just wouldn't feel right."

In the event you would drive through the light and generally regard yourself as above reification and internalization, think again. You have reified and internalized countless rules. How would you feel about having live ants and cockroaches for dinner tonight, for example? Are you willing to give them a try? How do you feel about murder, rape, and child abuse? Are you pretty casual about them, or do you feel they are *really wrong?* If you think about it, you'll find you feel pretty strongly about a lot of our rules.

All of this notwithstanding, sociology is also the study of how we *break* the rules. Some people use bad grammar and pour their soup on the floor, not to mention drive too fast, steal, fix prices, commit murder, and everything in between. Although this may seem like the study of "bad people," beware.

First, the rules of society are so extensive and complex that no one can possibly keep them all. For example, there is probably a street near you that is posted with a twenty-five-mile-per-hour speed limit; that's certainly a rule. And yet if you drive twenty-five miles an hour on that street during rush hour, you may discover you're breaking another rule. Your clue will be the honking horns and shaking fists.

Beyond the impossibility of obeying all the rules, you and I might agree that some of them ought to be broken. Consider the rule, in force a few years ago, that black people had to sit in the back of the bus in some parts of America. The people who finally broke that rule are considered heroes today. By the same token, you might disagree with rules that women can't fix carburetors, that men shouldn't cry, or that professors always know better than students.

The study of how people break the rules is closely related to the study of how the rules *change* over time. Although we are always living in a sea of rules, and many seem to last forever, it is also true that the rules of society are always in a process of change. Rules pertaining to hemlines, hair length, and political views operate a little like yo-yos. Others seem to change in only one direction.

Sociology, then, is an examination of the rules that govern our living together: what they are, how they arise, and how they change. Sociology, however, is a special approach to the rules of social life. As we'll see, there are other approaches.

A Science of Society

Sociology shouldn't be confused with social philosophy. It is not a point of view about the way things ought to be. Rather, sociology deals with the way things *are*. Moreover, sociology is more than just an *opinion* about the way things are.

Sociology is a science of social life. Like other sciences, sociology has a **logical/empirical** basis. This means that, to be accepted, assertions must (1) make sense and (2) correspond to the facts. In this sense, sociology can be characterized by a current buzzword: *critical thinking*. The simple fact is that most of us, most of the time, are uncritical in our thinking. Much of the time we simply believe what we read or hear. Or, when we disagree, we do so on the basis of ideological points of view and prejudices that are not very well thought out.

Suppose you were talking with a friend about the value of going to college. Your friend disagrees: "College is a waste of time. You should get a head start in the job market instead. Most of today's millionaires never went to college, and there are plenty of college graduates pumping gas or out of work altogether." That's the kind of thing people sometimes say, and it can be convincing—especially if it's said with conviction. But does it stand up to logical and empirical testing?

Logically, it doesn't seem to make much sense, since a college education would seem to give a person access to high-paying occupations not open to people with less education.

How does the assertion stack up empirically? Table 1.1 lists the median incomes of families headed by individuals of different educational levels as reported by the Census Bureau.

There isn't any scientific support for the assertion that education is a worthless financial investment—even though there are some individual exceptions to the rule.

It's important to recognize that human beings generally have opinions about everything. We're going to look at this more fully in Chapter 2, but it needs saying here that the people you deal with every day have a tendency to express opinions about the way things are—and what they say isn't always so. Consequently, you need to protect yourself from false information. That's what critical thinking is all about, and sociology provides some powerful critical-thinking tools.

Table 1.1
Median Income by Educational Level

Educational Level	*Median Income*
Less than 8 years	$ 9,221
8 years	11,811
1–3 years high school	13,705
4 years high school	20,800
1–3 years college	24,606
4 years college or more	34,709

Source: U.S. Bureau of the Census, *Statistical Abstract of the United States,* Washington, D.C.: U.S. Government Printing Office, 1985, p. 443.

I hope these few examples will indicate that sociology is not just something you might study in college and never think about again. Sociology deals with powerful issues that determine the quality of your life. Understanding sociology can empower you to be a more effective participant in the social affairs around you whether you are a conscious player or not. Marriage, employment, prejudice, crime, and politics are only a few of the areas of social life that can be importantly affected by your ability to engage in sociological reasoning.

Let's look at the twin foundations of critical thinking and of science in more detail—seeing how they apply to sociology in particular. Let's start with the determination of *facts*.

The Search for Facts: Research

At his August 5, 1985, news conference, then-President Ronald Reagan noted the fortieth anniversary of the Hiroshima bombing in the context of an unabated nuclear arms race and increasing popular demands for a freeze on the manufacture and deployment of nuclear weapons. In defense of his administration's military policies, the president suggested that our store of nuclear weapons, coupled with the horrible example of Hiroshima, "is a deterrent that kept us at peace for the longest stretch we've ever known, 40 years of peace."

Regardless of how you feel about the nuclear arms race—whether we should be in it and who's winning—the argument that our nuclear arsenal had at least kept us at peace was a compelling one. In my review of the media and public comments following the news conference, people disagreed on whether the nuclear deterrent was the reason for the forty years of peace. Yet to my amazement, no one questioned the empirical accuracy of the assertion. A simple historical review of the period from 1945 to 1985 turned up the Korean War from 1950 to 1953 and the extensive and expensive war in Vietnam, which claimed 47,318 American lives between 1959 and 1973. This has to have been the bloodiest forty years of peace in our national history.

Much of what you read and hear simply doesn't correspond to the empirical facts, but the inaccuracy gets overlooked in the heat of rhetoric. For example, when the newspaper *USA Today* focused its attention on the national budget deficit, reporters interviewed seven men and women "in the street" to get their opinions. One of those interviewed, Glen Cemer, a retired photographer from Battle Creek, Michigan, responded as follows: "One way to lower the deficit would be to reduce the amount of foreign aid we provide to other countries—it's not winning any friends for us. Let's use that money to lower the deficit."[3]

This comment is often heard when people discuss ways to reduce the federal budget deficit. On the one side, some people, like Mr. Cemer, can point to instances of foreign aid being wasted in various ways and cases where countries we have aided have subsequently turned against us. On the other side, people who support foreign aid point to positive accomplish-

---

3.   *USA Today,* September 25, 1985, p. 10A.

ments against a general backdrop of our moral urge to help others less fortunate than ourselves.

Regardless of how you happen to feel about foreign aid in general, it is possible to determine, empirically, whether reducing foreign aid would be an effective way of reducing the federal budget deficit. In 1990, the United States gave a total of $10.8 billion in economic (nonmilitary) aid to other countries. Our budget deficit that year was $238 billion. Canceling our foreign aid program altogether, therefore, would have reduced the deficit by only 4.5 percent. Whereas the budget deficit is sometimes expressed on a per capita basis, canceling foreign aid in 1990 would have reduced the 1990 deficit of each American from $956.98 to $913.55—not much of a reduction.[4]

One of the things I hope you'll gain from this introduction to critical thinking is an ability to balance what you read and hear in your day-to-day life against empirical evidence. That's one of the keystones of sociology.

Over the past 150 years, sociologists have developed a number of research methods to assist in determining the facts of social life. **Survey research**, for example, involves the use of questionnaires to collect information. Sometimes it's appropriate to have people fill out the questionnaires themselves; other times interviewers read the questions and take down the answers. In recent years, telephone interviewing has become increasingly popular and effective. In particular, computers have been adapted to support the interviewing process: selecting telephone numbers at random, showing interviewers what questions to ask, and recording the answers given for immediate analysis.

Usually, surveys are conducted among samples of people, selected in such a way that a few hundred or thousand respondents can provide an accurate assessment of much larger

---

4. U.S. Bureau of the Census, *Statistical Abstract of the United States, 1992,* Washington, D.C.: Government Printing Office, 1992, pp. 8, 279, 794.

populations. For example, a political poll among a properly selected sample of 1,600 voters can provide an accurate estimate of how all 100 million American voters feel: accurate within 2.5 percentage points. I know that may be hard to believe, but sociologists are putting the logic and techniques of survey sampling to effective use every day.

Sometimes, laboratory **experiments** are a more effective method of getting at the facts of social life. This technique is commonly used in the study of small group dynamics, for example. As a practical example, some sociologists have used this technique to study the dynamics of jury deliberations. Hiring subjects to serve as pseudo-jurors, the researchers can then observe their subjects argue their way to a verdict.

A ready resource for sociological analysis exists in the great masses of data regularly compiled by governmental agencies and various groups in the private sector. As in the example of education and income that we examined earlier in the chapter, it's often possible to find appropriate answers without the time and expense of conducting a survey, experiment, or comparable research project. Even in this situation, however, you need to exercise critical thinking, since the facts never really "speak for themselves." Here's an example.

When Ralph Nader published *Unsafe at Any Speed*[5] in 1965, he launched a consumer rights movement that has subsequently touched most aspects of American life. In the realm of auto safety, numerous changes have occurred in the past twenty years: safer auto designs, mandatory seat belts, and reduced highway speed limits, to name a few.

Although everyone is in favor of highway safety, the kinds of changes I've mentioned have been very controversial. Some people argue that the various changes have made driving safer; others disagree. This is a good place for empirical examination.

Let's see what has happened to the number of people killed on the nation's highways during this period of activity.

---

5. Ralph Nader, *Unsafe at Any Speed,* New York: Grossman, 1965.

Figure 1.1
Highway Deaths per 100 Million Miles Driven: 1965–1984

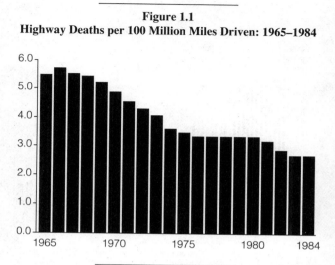

Source: Data taken from U.S. Bureau of the Census publications: *Statistical Abstract of the United States,* Washington, D.C.: U.S. Government Printing Office, 1981, p. 622; 1982–1983, p. 615; 1984, p. 615; 1985, p. 599; and *Historical Statistics of the United States: Colonial Times to 1970,* Washington, D.C.: U.S. Government Printing Office, 1975, pp. 719–720.

Before we simply check on the *number* of people killed each year, we should recognize that our population—including the number of drivers—has grown each year. More people have been driving more cars more miles each year. To make a fair comparison, therefore, we need to take account of those increases.

Figure 1.1 indicates the number of traffic deaths per 100 million miles driven each year from 1965 to 1984. This *rate* takes account of the increasing numbers of drivers, cars, and miles, and it gives us good grounds for observing trends in highway deaths over the period in question.

As a general rule of thumb, it is usually a bad idea to evaluate any time-series data without seeing them within the context of a longer period of time. Although it certainly looks as though

Figure 1.2
Highway Deaths per 100 Million Miles Driven: 1925–1984

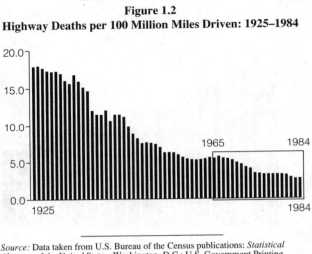

Source: Data taken from U.S. Bureau of the Census publications: *Statistical Abstract of the United States,* Washington, D.C.: U.S. Government Printing Office, 1981, p. 622; 1982–1983, p. 615; 1984, p. 615; 1985, p. 599; and *Historical Statistics of the United States: Colonial Times to 1970,* Washington, D.C.: U.S. Government Printing Office, 1975, pp. 719–720.

highway death rates have been dropping since 1965, what was the trend like earlier? Figure 1.2 answers that question.

What does this new graph show us? The most striking observation is that the highway death rate has been decreasing pretty steadily since 1925, the first year statistics are available. In fact, the decrease from 1965 to 1984 is dwarfed by the reductions prior to that.

The observed decline in highway deaths between 1965 and 1984, then, does not prove that the consumer activism of that period was the cause. Do the data prove that the consumer activism had no effect? Not at all. We cannot assume from the data of 1925–1965 that the long-term decline in highway deaths would have continued—without the speed laws, seat belts, etc.

There are legitimate reasons for suggesting that the long-term decline would not have continued. First, as the death rate approaches zero, it begins to evidence a *floor-effect*. It can never go below zero, and it's unlikely that it will ever even reach zero: there will probably be some highway deaths as long as there are highways and drivers. It is quite possible that the highway death rate was about to level off at five to six per hundred million miles driven.

Second, a closer inspection of the long-term graph (Figure 1.2) shows a small but steady increase in the death rate from 1961 through 1964. Although that's too short a period to draw long-term conclusions from, it is certainly possible that the death rate was beginning to rise again.

If either of these possibilities was actually true, the data would suggest the consumer activism beginning around 1965 did, in fact, have an impact. At this point, the data are inconclusive. I have pursued this example for two reasons. First, I want to demonstrate that looking at the facts requires some sophistication and thoughtfulness. Often, you need to look below the surface to see what's going on.

Second, I want you to see that facts alone are seldom enough. Sociology is called logical/empirical because facts and reason go hand in hand. You need both. Let's turn now to the matter of reasoning.

The Role of Reason: Theory

In the midst of national concern over the persistence of hunger in America, then-President Reagan appointed a thirteen-member Task Force on Food Assistance to determine the nature and extent of the problem and to recommend government action. One of the panel's members, Dr. George

Graham, attracted considerable attention by indicating his belief that the problem of malnutrition among children had been exaggerated—especially with regard to black children. He was quoted as stating:

> **National data show that black children are now taller than white children—obviously, they must be getting more to eat. . . . If you think that blacks as a group are undernourished, look around at the black athletes on television—they're a pretty hefty bunch.[6]**

As you can see, Dr. Graham cited two pieces of evidence—one statistical and one impressionistic—to support his thesis that black Americans are no more likely to suffer malnutrition than other Americans. But take a minute to consider these assertions logically.

First, Graham said black children are taller, on the average, than white children. But, even if this is true, does it mean that black children "must be getting more to eat"? What other factors might affect height? If you said "genetics," go to the head of the class. Racial and ethnic groups differ in average height. Consider the African Tutsi tribe, for example. These poor cattle-herders of Burundi and Rwanda, noted for their spectacular height, are commonly over seven feet tall. It would be absurd to conclude that they must eat more than Americans just because they are taller.

Second, what about the black athletes on television? Does their heftiness mean that black Americans as a whole must be getting a lot to eat? In this instance, you might have asked whether the black athletes are typical of all black Americans. Of course not. They are no more typical of blacks in general than white athletes are typical of all whites. Otherwise, we would need to look at Japanese sumo wrestlers and conclude that the Japanese people in general must be bigger and better fed than Americans.

---

6. "Big Changes Made in Hunger Report," *San Francisco Chronicle*, December 30, 1983, p. 10.

As another example of the function of logical reasoning in everyday discourse, let's consider the broad public concern over acquired immune deficiency syndrome (AIDS). This extremely deadly disease has been disproportionately concentrated among homosexuals in America, especially during the early years of the epidemic, thereby coloring the public and political response to the disease. American gays and those sympathetic to the gay community have frequently charged that research on the disease has been slow in coming, due to prejudice against homosexuality.

Regardless of whether the charge of prejudice is true in general, there is no question that some people feel AIDS is a fitting punishment for gays. Some have even argued that AIDS is God's way of expressing divine displeasure with homosexuality. The discovery that AIDS is also commonly spread by contaminated needles of drug users and that it is also spread by prostitutes only confirms the theory of divine retribution.

Evangelist Don Boys, writing a guest editorial for *USA Today,* expressed his feelings this way: "The AIDS epidemic indicates that national morality has broken its mooring and drifted into a miasmic swamp, producing disease, degeneracy, and death. . . . God's plan is for each man to have one woman—his wife—for a lifetime, and be faithful to her. God's plan still works." [7]

Although sociology has no direct pipeline to divine will, critical thinking can shed some light on this assertion. To begin, AIDS has also been disproportionately common among Haitians in America. The divine retribution theory would need to include God's displeasure with them. The susceptibility of hemophiliacs and others needing blood transfusions further complicates the matter. The fact that AIDS is primarily transmitted through heterosexual sex in Africa muddies the water still further.

---

7. Don Boys, "Don't Spend Money; Stop Sinners Instead," *USA Today,* October 7, 1985, p. 6A.

But let's put the divine retribution theory to its acid test. If AIDS is a sign of God's preference for lifestyles, then lesbians must be the most favored of all, for AIDS is least common among lesbian women! In short, the divine retribution theory, as generally conceived, simply doesn't hold up under logical scrutiny.

Logical reasoning, then, is another key component in critical thinking. As a science, sociology is consciously and deliberately committed to logical reasoning, and, as the above examples indicate, such a concern has real, practical implications. In part, this involves the development of **theories.** To understand the function of theory, we need to consider a few of its elements.

First, **concepts** are the mental images we use to bring order to the mass of specific experiences we have. Basically, we form concepts in order to group similar things together and to distinguish dissimilar ones. For example, the concept "human being" groups nearly 6 billion creatures into a single category and distinguishes them from billions and billions of other creatures. "Male" and "female" divide the 6 billion human beings into two categories.

Other concepts that distinguish different kinds of human beings include "infants," "Ethiopians," "mail-carriers," "college students," and so forth. All of us use concepts like these every day. We also commonly use concepts that do not necessarily distinguish different kinds of people. We may speak of "anxiety" and "alienation," for example. "Crime" is a concept; so is "unemployment."

Sociologists work with many of the same concepts that people commonly use in everyday life, but, as scientists, sociologists are more deliberate in their use of concepts. For example, sociologists are particularly concerned with the precise definitions of concepts such as those we've been considering. How do we decide, for example, if a person is unemployed? We probably wouldn't say a newborn baby was unemployed, but what about a wealthy playboy who never had a job and doesn't plan on ever having one? Specifying exactly what we will mean by the concepts we use is a fundamental part of sociology.

Equally important, sociologists seek to discover order among concepts. The concepts "male" and "female" are often referred to as **attributes** of individuals. The **variable** "gender" is a concept that brings together those two attributes. The variable "religion" brings order to such attributes as "Protestant," "Catholic," "Jew," and "Hindu."

It could be said that sociology is devoted to discovering relationships among the variables that distinguish different kinds of people. The earlier example of education and income illustrates that. The relationships to be explored are limitless. Are married people happier than single people? Are conservatives more prejudiced than liberals? Do broken homes produce juvenile delinquency?

Sociological theory goes a step beyond the questions of relationships between pairs of variables, however. Ultimately, we seek to develop a more or less comprehensive picture of the interrelations among a great many variables—ideally to understand the functioning of social life in general.

As a basis for such comprehensive understanding, sociologists begin with overarching models or **paradigms** of social life. A physical example may help introduce the idea of paradigms.

Observe the muscular bicep in Figure 1.3. This is something I know you have seen more than once.

A physiologist looking at this flexed bicep might see something quite different. Notice how the muscle could look like a network of tendons to a physiologist (Figure 1.4).

A cellular biologist, on the other hand, might see the tissue of the bicep as a system of cells, each with its nucleus and other cellular particles, as in Figure 1.5.

A molecular biologist would take an even closer look, however: seeing the interlocking system of various molecules (Figure 1.6).

Figure 1.3
Bicep

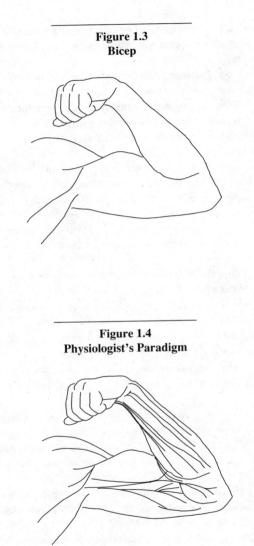

Figure 1.4
Physiologist's Paradigm

Figure 1.5
Cellular Biologist's Paradigm

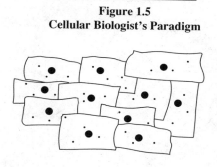

Figure 1.6
Molecular Biologist's Paradigm

Finally, an atomic physicist would look into each of the molecules to examine the component atoms, each built from protons, electrons, and other subatomic particles, as Figure 1.7 illustrates.

Each of these diagrams illustrates a different paradigm that might be used in "looking" at the flexed bicep. None of the paradigms is better than the other; each simply offers a different perspective that might be more or less useful for a given purpose. That's the nature of paradigms.

Although these illustrations have moved in a particular direction—more and more microscopic—that's only one possible line of variation in paradigms in this case. A different paradigm might see the flexed bicep as a sign of virility, a

Figure 1.7
Atomic Physicist's Paradigm

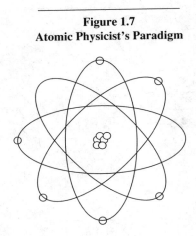

Figure 1.8
A Macho Paradigm

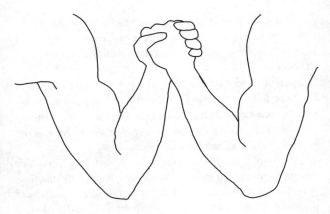

threat of violence, or part of a bodybuilders' contest. Perhaps you can think of other paradigms that would portray the flexed bicep in other ways. (Figure 1.8)

There are three major paradigms commonly used in modern sociology. Some sociologists find it most useful to regard social life as a multitude of **interactions** among people. Consider a conversation, for example. Let's assume that you and I have met for the first time, waiting in the dentist's office. There are no magazines, not even old ones. All we have is each other.

After staring at the ceiling awhile, I break the ice by saying, "Waiting for the dentist?" You confirm my brilliant insight. "Been here before?" you ask. I reply that I come for a check-up every six months, that I floss every day, and that I have three gold crowns. Pretty soon we are engaged in an adventure of mutual discovery. Before long, we've discovered that we both grew up in Vermont, that we tend to vote for different political parties, and that we both own malamute dogs. When I say I'm a sociology professor, you say you took a course in sociology once. Now, I may begin trying to shape the conversation into a pattern appropriate to a professor–student exchange, and you try to get back to the topic of malamutes.

The **interactionist paradigm** in sociology focuses on social life as the process of give and take in which individuals come to grips with each other, forming common definitions of the situation they find themselves in. This paradigm is particularly useful in understanding the ways in which we create the rules of social life.

A very different paradigm, the **social systems paradigm**, or **functionalist paradigm**, tends to focus on the structure of social life. A group of people, as large as a whole society, can usefully be seen as constituting an integrated system in which each individual has a part to play.

Consider a symphony orchestra, for example. The violins have a particular job to do (or function to perform), the horns have a different job, and the conductor's job is different from any of the players. Taken together, all the members of the orchestra constitute a whole system that is greater than the sum of its parts.

A football team offers another illustration of how we might use the social systems paradigm. The quarterback has one set of functions, and the running backs, linemen, and receivers have different functions.

Shifting to a whole society, we might examine the functions played by groups and organizations such as factory workers, police, teachers, children, ministers, and so forth. This paradigm is particularly useful in understanding the organization of the rules of life we've created.

Finally, the **conflict paradigm** focuses on the competitive struggle among individuals and groups in society. Criminals conflict with police, workers with owners, students with teachers, Protestants with Catholics, blacks with whites, and so forth. Typically, these struggles grow out of the unequal distribution of quantities perceived as desirable and limited, such as money, property, prestige, and power. This paradigm is particularly useful in understanding how and why people break the rules and how the rules themselves change.

Be clear that none of these paradigms is intended as a complete statement of how social life operates. Rather, each offers a different point of view that reveals certain aspects of social life and conceals others. The obvious implication of this is the value of using more than one paradigm for the purpose of achieving an integrated, well-rounded view. Throughout the remainder of the book, we'll see ways in which these paradigms are useful windows on our world. In fact, if you were to learn no more sociology, you could make use of these three paradigms, just from what you already know about them. As you learn more, of course, they become more useful.

Sociological Questions and Answers

The common image of science is of scientists finding the answers to questions. I will conclude this chapter with a somewhat different view of science and of sociology in particular. Science is sometimes better at raising questions than at finding answers to them. It would be useful for you to regard science as an ongoing inquiry, recognizing that questions initiate new avenues of inquiry, while answers close them down.

Science makes an especially powerful contribution when it calls into question those things that "everyone knows." Everyone used to "know," for example, that blacks were inferior to whites and that women were inferior to men. As you'll see, sociological points of view very often raise questions about things everyone else thought had already been answered. Even when we discover new and seemingly better answers about something, it's important to hold them as tentative.

There is a particular **recursive** quality in human life that makes anything we know tentative. Whenever we learn something about ourselves, what we've learned may bring about changes—even to the extent of making what we learned no longer accurate.

Suppose we studied employment opportunities across the country, for example. When our study was complete, we listed the ten cities with the most jobs available. As soon as our findings became widely known, of course, a lot of unemployed people would move to those cities, and soon those cities would not have as many jobs available as before. This is the same thing that happens when a newspaper columnist identifies a local restaurant that has great food, low prices, and no waiting. The "no waiting" part probably won't last another twenty-four hours, and the other two characteristics may disappear, too. In sociology, anything we learn can change things, so no knowledge can be counted on to remain true. Thus, we need to keep asking questions.

Even more fundamentally, sociology deals with a number of questions that will never be answered fully. Who am I? What is a human being? Are we more the result of our genes or of our environment? Is order possible without restricting freedom? These are a few of the questions we'll be addressing in the rest of this book, and it is unlikely that they will ever be completely answered. As we'll see, however, it can be very useful to keep asking them anyway.

I point all this out to give you an appropriate context for your own inquiry into sociology. Although there are some facts about sociology that are worth learning, it is more important for you to learn to use sociology for your own ongoing critical thinking. If you were studying brain surgery or medieval history, you might not have an opportunity to use what you learned in your day-to-day life. Sociology is very different. Every day you will wake up into a sociological laboratory with a massive experiment underway. We're all subjects in the experiment, and you now have the opportunity to join the researchers.

The remainder of this book will introduce you to the multifaceted worlds of sociology. As an overall structure for our examination, we will start, in Chapter 2, with a look into perhaps the most intimate aspect of sociology, that of self-concept.

We'll be examining the age-old and still-unanswered question: who am I? As you'll see, your most fundamental notions of who you are as an individual are inseparable from the society in which you live. This concern, along with those pertaining to small-group interactions, is often labeled **microsociology.**

Beginning with personal identity in Chapter 2, our field of focus will grow ever wider with each successive chapter: **groups** in Chapter 3, **organizations** in Chapter 4, **institutions** in Chapter 5, and on to **society** and **culture** in Chapter 6—having progressed fully into the realm of **macrosociology**.

Finally, having dealt with some broad-ranging societal issues in Chapters 7–9, we'll conclude with an examination of some global matters. My intention in all this is to let you see clearly some of the ways in which your most intimately individual being is inseparable from life on the planet as a whole. That's a part of what the sociological spirit has in store for you, and I think you'll find it fascinating.

CHAPTER 2

IDENTITY

In the early 1970s, a team of psychologists at Stanford University conducted an ill-fated experiment that said more about human beings than has generally been recognized.[1] Under the direction of Dr. Philip Zimbardo, the Stanford Prison Experiment was designed for the purpose of learning something about the social psychology of prisons. To accomplish this, Zimbardo recruited Stanford students to act out the parts of prisoners or guards in a physical simulation of a prison in the basement of a campus building.

Zimbardo was careful to screen the student volunteers to ensure that none suffered from psychological problems that might endanger either themselves or the experiment. All were judged "normal"; Zimbardo would later describe them as "the cream of the crop of this generation." The two dozen students were assigned by the flip of a coin to serve as either prisoners or guards in the experiment. Each student was paid fifteen dollars a day for participating in what was planned as a two-week experiment.

The experiment began smoothly enough. The "prisoners" were picked up unexpectedly at their homes, taken away in handcuffs for booking, and eventually assigned to their "cells." The "guards" were given the chore of developing their own rules and procedures for maintaining law and order. Prisoners

---

1. See Philip G. Zimbardo, "Pathology of Imprisonment," *Society,* 9 (April, 1972), pp. 4–8.

were ushered to and from the bathroom, and their meals were brought to them in their cells. It could have turned out to be a pretty boring activity, but it didn't turn out that way.

Early in the experiment, Zimbardo found the guards were taking their jobs more seriously than had been expected. There was a tendency for them to become authoritarian in their dealings with the prisoners—and some even showed sadistic tendencies. Although not all the guards became cruel and tyrannical, Zimbardo noted that none of the "good" guards ever stepped in to interfere with the cruelty of the "bad" guards.

The prisoners were becoming realistic, too. Some became troublesome, verbally abusing the guards and threatening to damage the sparse furnishings of their cells. Other prisoners evolved in just the opposite direction, meekly following all the orders of their guards.

Whereas the prisoners had begun with some degree of good-natured solidarity as a group, the guards succeeded in breaking that down, creating a feeling of worthlessness among individual prisoners as well. At one point, for example, the guards decided to put one of the prisoners in solitary confinement because he refused to eat. Then they offered the other prisoners a choice: if they volunteered to give up their own blankets for the night, the troublemaker would be brought back to his cell instead of spending the night in the small closet established as "the hole." All the prisoners voted to keep their blankets.

Three prisoners had to be released during the first four days: the experiment had become too traumatic for them. The other prisoners eventually begged to be let out of the experiment; most said they would forgo the money they had already earned.

Zimbardo was astounded at the changes the volunteers were undergoing, but he was also struck by his own reactions to the experiment. Zimbardo had established himself in the role of prison superintendent, and when he learned of a rumor that a campus fraternity was planning a late-night raid on the prison to spring the prisoners, Zimbardo panicked! At first, he began calling local police authorities to see if they would lock up his prisoners for safekeeping. When they refused ("You want us to do what?!"), Zimbardo secretly transferred the prisoners to another room on campus, where he kept them handcuffed to each other through the night.

When Zimbardo saw how fully he and the students had fallen into their roles, he recognized that the experiment had become something ominously more than intended. Consequently, he terminated the experiment early—*after only six days!* Moreover, Zimbardo felt it appropriate to conduct counseling sessions with the participants to undo the social learning that had occurred during the experiment.

More than anything else, I think, the Stanford Prison Experiment illustrates the extent to which our behavior is conditioned by the situations we find ourselves in, by the roles we are called upon to play. The student volunteers at Stanford received no real instructions in how to be prisoners or guards. And although we can assume that each came to the experiment equipped with some amount of folklore from television and movies, the realism of their behavior quickly surpassed that informal preparation. There was simply something about being in a prison—even a simulated one—that brought out the same kinds of behavior exhibited by real prisoners and real guards in real prisons.

The Stanford Prison Experiment touches on an even more fundamental and profound issue—one that I will address in this chapter. It raises questions about human identity. Who are you? Who am I? What is a human being?

Who Are You?

For the next few minutes, I would like you to explore with me the question "Who am I?" You have a sense that you *are*. But who or what is it that you are? I think the answers you discover in this inquiry will be of more than academic interest to you.

In 1954, Manford Kuhn and Thomas McPartland devised a questionnaire to deal with this issue.[2] The model of simplicity, the questionnaire asked respondents (usually college students) to write down twenty answers to the question: "Who am I?" In the years since, untold thousands of students have completed the Twenty-Statement Test (TST), as it has become known. Moreover, researchers have discovered certain patterns in the answers people give.

Take a minute to consider how you would answer the test. You might even want to take just a few more minutes and write down twenty answers.

If you are at all typical, the first answers you would give to the TST could be described as *statuses* that you occupy. You might have answered "student," for example. Or perhaps you said "man" or "woman." Some people answer by giving their race or religion. Some give their age as an answer. All these are characteristics that everyone would probably apply to the person in question.

Typically, respondents go a step further and describe themselves in ways that might be more of an opinion: "I am a happy person," "I am overweight," "I am conservative." Although others might disagree with us in some of the ways we characterize ourselves, these latest answers also can be seen as examples of statuses.

Statuses and Roles

Status is a basic concept in sociology. It refers to a *position* or *location* that a person can occupy within society or within a smaller social grouping. Mother, plumber, sophomore, and safecracker are all examples of statuses. In the discussion of social systems theory in Chapter 1, you'll recall my saying that each member of a society has a function to play. More

---

2. Manford Kuhn and Thomas McPartland, "An Empirical Investigation of Self-Attitudes," *American Sociological Review,* February 1954, pp. 68–76.

Figure 2.1
Statuses as Masks

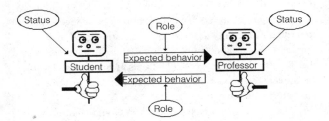

accurately, we should say that statuses have functions to serve, and these are commonly called the **roles** associated with and expected of a status. If you occupy the status of mother, for example, certain kinds of behavior are expected of you that would not be expected of a plumber.

In review, then, a status is a position in society and roles are the behavior expectations associated with a status. Status is a matter of being, role a matter of doing. We occupy statuses and perform roles.

For the most part, statuses come in pairs, and the roles expected of us are in relation to other specific statuses. Each status is like a mask we hold up when we interact with others (see Figure 2.1). Thus, if you and I met, we might hold up the status/masks "student" and "professor."

Even though there are no rigid, totally-agreed-upon rules for the behavior between professors and students, all of us who have occupied either or both statuses have some shared understandings about what's expected. At the very least, we know those expectations are quite different from what would be expected between a mother and her son, a prisoner and a guard, or a salesclerk and a shopper. If you and I were to meet for the first time, we'd want to discover the statuses we occupy in relation to each other—and we'd behave accordingly.

Figure 2.2
We Each Occupy Several Statuses

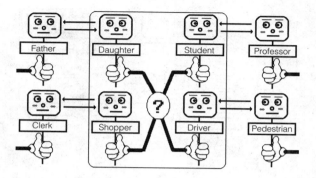

In real life, of course, each of us occupies a number of statuses, acting out the roles of different statuses in different situations. Thus, in some situations, it might be appropriate for you to act out the role of son or daughter; at other times "student" might be a more appropriate role to play. At still other times, you act through the statuses of driver, shopper, citizen, or friend. Figure 2.2 gives a simple illustration of the set of statuses an individual person might occupy.

Notice the question mark in the center of the diagram. I've put it in to represent whoever or whatever is holding the several masks, occupying the various social statuses. The question mark, in some sense, represents *you,* and I'd like you to keep asking who you are as we continue.

Each preceding diagram is, of course, overly simplified. For example, each status that you occupy has role relationships with numerous other statuses. Thus, a college student is subject to expectations about behavior in relationships with professors, with other students, with deans, with coaches, with exam proctors, and so forth. Figure 2.3 illustrates that aspect of roles and statuses.

Figure 2.3
Each Status Relates to Many Other Statuses

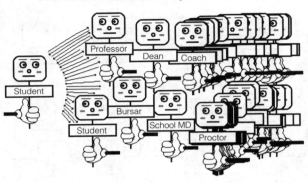

Putting together what we have examined so far, we see that each individual occupies countless statuses, each of which has role expectations in relationship with several other statuses. Take a minute to think about all the statuses you occupy and all the statuses each of those relates to, and you may be justifiably impressed by how much you have learned about what it means to be a human being in organized society. Figure 2.4 gives a greatly oversimplified picture of the playing field for being human.

An important part of growing up and of socialization involves your learning the statuses you occupy, the role expectations associated with those statuses, and your relationships with people occupying other statuses. This process never ends. No sooner do you master the rules of being a child than you find yourself playing teenager. Once you get that down pat, you discover you've moved on to "responsible adult." Eventually, you may hear people saying, "Act your age, you old coot!" No matter how you feel inside, the statuses you occupy—such as age level—determine how others expect you to act.

The never-ending socialization process is not limited to age changes. If you marry, you will discover (despite what your mate

Figure 2.4
The Playing Field for Being Human

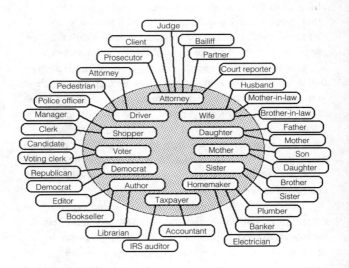

may have assured you) that you've inherited a radically new set of role expectations. Become a parent and they change again—even more radically, perhaps. Or suppose you move from medical student to intern to surgeon to chief of surgery to Surgeon General of the United States. Each of those shifts brings new and different expectations for your behavior. Whatever career you choose for yourself, your statuses and associated role expectations will change over time. Even if you wash dishes in a diner all your life, you'll move from novice to veteran.

The Three Paradigms

By now, you may have gathered that sociologists regard status and role as rather important elements of society. Notice how they figure within the three theoretical paradigms we examined in Chapter 1: the interactionist, social systems, and conflict paradigms.

We've already seen how our *interactions* are mediated through the statuses we occupy. Take a minute to try imagining what it would be like to interact with someone in total isolation from the statuses you occupy. You couldn't see each other because what you learned about each other's sex, age, and race would structure the situation. You couldn't talk over the phone because your voices would give clues about statuses. Perhaps you could carry out a conversation with each other, typing on computer terminals, but you can bet that much of your attention would be devoted to learning something about the other person. It's no surprise that we so often resort to "What do you do?" or "Tell me something about yourself" when we meet new people. In college, of course, one of the most clichéd lines of any social mixer is "What's your major?" In short, social interaction is simply inconceivable in the absence of statuses.

Statuses are no less fundamental to the *social systems* paradigm. All organizations and society itself are structured and operate on the basis of social statuses. A symphony orchestra, for example, is organized in terms of statuses, not people. Even though a particular musician may stand out as a virtuoso at a given time, the orchestra doesn't disband when that person retires or dies. Someone else is found to occupy the vacated status. By the same token, music is composed to be performed by statuses: violinists, oboists, cellists. Even when a piece of music is composed with a particular virtuoso in mind, it can still be performed by others who play the same instrument.

Corporate structures and factories also offer excellent illustrations of the place of statuses in the social systems paradigm. The assembly line, for example, can only function through the role expectations of welders, joiners, supervisors, managers, and so forth. The corporate table of organization is an organization of statuses, not of people. In sum, then, social systems could not exist without statuses.

Finally, *conflict* theory is also fundamentally grounded in social statuses. The "conflict" in conflict theory occurs

Figure 2.5
The Place of "Status" in Theoretical Paradigms

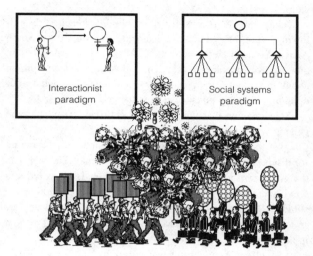

Conflict theory paradigm

between the occupants of different statuses and status categories: workers vs. owners, peasants vs. landowners, students vs. teachers, gang vs. gang, nation vs. nation. Status was essential to Marx's concepts of class struggle, class consciousness, and the like. Conflicts between specific individuals are of little or no interest to conflict theorists unless they represent a struggle between *kinds* of people.

Figure 2.5 illustrates the place of status within the three sociological paradigms.

Opinions as Identity

The opinions we hold provide the basis for a special kind of status—and thus another basis for identity. It may not be

apparent to you just how important it is for you to have opinions. Years ago, the Beatles sang of the "nowhere man" who
didn't have a point of view. There's a lot of truth in that observation.

Take a minute to consider your opinion on each of the following: abortion, capital punishment, war on drugs, South
Africa, Somalia, Jerry Falwell, Bill Clinton, overpopulation,
gun control, welfare. My hunch is that you had an opinion
about each one. For my purposes, it matters less what your
opinion was in any given case than that you had one.

Now imagine that you were at a gathering of people you
generally respect and whom you want to think well of you.
Imagine that any one of the topics mentioned above came up
for discussion: capital punishment, for example. The others in
the group turn to you to learn your opinion. How would you
feel if you were to say you didn't really have an opinion on the
subject? How do you suppose the others would feel about you?
In an important sense, you wouldn't fit in. You'd be nowhere
in the matter, almost as though you didn't exist.

A couple of years ago, I was struck by the overwhelming
need I felt to have an opinion on every controversial issue.
Specifically, there was an international controversy concerning
two test-tube babies in Australia. A wealthy couple there had
been unable to conceive, so they resorted to *in vitro* fertiliza-
tion. An egg had been taken from the woman's womb and had
been fertilized by sperm taken from the husband. Two fertilized eggs were alive and well, ready for implantation back in
the woman's womb.

Then tragedy struck. The couple was killed in an accident,
leaving the two fertilized eggs orphaned even prior to birth.
The controversial question was this: should the two eggs be
destroyed, or should they be implanted in some other woman
and brought to birth? The controversy was probably fueled by
the fact that a substantial inheritance was involved.

What do you think? Should the eggs have been destroyed
or brought to birth? The more I thought about it, the more frus-

trated I became at not being able to decide how I felt. I
couldn't decide what the Australian authorities should do. If
you are anything like me, you may feel the need to have an
opinion in the matter.

Eventually, however, I was struck by the stupidity of my
dilemma. I had no expertise to bring to the issue, so my opinion wouldn't have any particular value. In fact, who would
really care what I thought? Certainly any opinion I might
express would have no impact on what happened in
Australia. Even recognizing all that, however, I still felt a little incomplete that I didn't have an opinion about what
should be done.

That experience drew my attention to the extent to which
our identities get wrapped up in our opinions. Take a minute to
see if it isn't so for you. You might want to recall the last time
you met someone new and engaged in the interactional process
of "getting to know each other." If you can recall such an
episode, I'll bet your interaction was largely an exchange of
opinions—discovering which ones you agreed on. The search
for agreement may have involved politics, sports, music, or
any number of other topics.

By identifying ourselves with particular opinions, we establish more statuses that we occupy. Say you are totally committed to an immediate halt to limits on abortion, and you
announce your occupancy of the status of "pro-choice."
Declare that welfare and all other social programs should be
abolished and that corporate taxes should be reduced, and you
establish your credentials in the status of "conservative." At
the very least, any opinion you express earns you the status of
"someone who believes that . . ."

The statuses you claim for yourself through the opinions
you hold and express have a powerful impact on your participation in and experience of life. They go a long way toward
determining how others deal with you: who will talk to you, be
your friend, ask your advice. Your opinion statuses strongly
influence who you select as a mate and who selects you.

Who Are You, Really?

All the preceding comments show how implicitly our identities are a function of the statuses we occupy. But who are *you* in relation to the statuses you occupy? I know you have an experience of *being* that is separate from those statuses. Suppose, for example, that one of your statuses is that of "college student." That would mean people often act toward you in ways appropriate to dealing with college students and that you often behave like a college student. Probably if you met a stranger who asked, "What do you do?" you'd say, "I'm a college student."

Yet, if you were to stop being a college student—regardless of whether you graduated or flunked out or just quit—you would still have a sense of being *you*—the same you as before. Who you are, then, is something other than your student status.

Suppose you switched from being a Republican to being a Democrat, or suppose, born and raised a Roman Catholic, you became a Buddhist. Or suppose you left the United States and became a citizen of Ethiopia. None of these changes of status—no matter how radical—would cause you to doubt that you were still the same you. You'd still be the same you if you changed your name. Who you are, then, is more fundamental than your political affiliation, religion, citizenship, or name.

Each of us has an implicit sense of *persistence* or *continuity* that can survive changes in everything you might point to in answering the question "Who are you?" You have an implicit sense that you are that same one who was born on your birthdate years ago, and yet, you don't look anything like that newborn baby. The cells of your body are replaced every seven years or so. Your current personality is nothing like whatever might have passed for personality in that baby, and your personality is probably still changing. Your opinions, knowledge, and memories keep changing, and yet you still have the sense of being the same you across all those changes.

Sociology is not designed to address who you *really* are behind your statuses and roles. As we have seen, the three sociological paradigms are inextricably wedded to the concept of status, so they cannot not reveal whatever it is that occupies those statuses.

At the same time, however, sociology can reveal how statuses operate in your life as well as in society at large. This can be very useful to you personally, not just scholastically. The statuses you occupy are fundamental to your experience of life—so fundamental, in fact, that they can interfere with your experience of who you really are.

Sometimes, we become so identified with the statuses we occupy that we suffer personally when those statuses are disrupted or threatened. Essentially, we lose sight of the fact that a particular status is *only* a status and begin thinking it represents the totality of who we are. Think of all the students who committed suicide because they flunked out of school or failed an important exam. Other people commit suicide when they fail in a career position. None of this could happen if we remembered that our statuses are simply masks we put on for the purpose of participating in society. If one mask didn't work out, we'd simply discard it and try another. Instead, we tend to start thinking the masks are us and vice versa.

Recall the last time someone said something nasty about all members of your race, gender, or similar status category. Did it make you angry? Did you say something nasty in return? Did you complain about the comment to someone else later on? In short, did you get hooked by the barb of the comment? If you didn't get hooked the last time someone said something nasty, recall a time when you did get hooked.

Looked at rationally, nasty remarks about "all blacks," "all women," "all Protestants," "all vegetarians," "all Republicans," or the like are pretty stupid. It's hard to think of anything meaningful you could say that would truly apply to all members of a large social category, nor is anyone in a position to determine the truth of any such statement if it were actually true. In short, only jerks make nasty remarks about whole social categories. Rationally, then, such comments and commentators should probably just be ignored as unworthy of serious consideration.

And yet, we do often take them seriously. We take what they say as personal attacks on our inner being rather than as stupid remarks about statuses we happen to occupy. Go tell a psychology professor that you think psychology is completely worthless and see what I mean. Or tell Harvard alumni that all Yale graduates are smarter. Better yet, don't do that to other people—just notice how many of your statuses you'd fight for.

A great deal of the prejudice, hatred, and war in the world is grounded in people identifying with the statuses they occupy and insisting that theirs are somehow superior to those occupied by others. Recognizing this trap in your own life may offer you some degree of sanity and protection from needless pain. It may even equip you to make a dent in the problem as it exists around you.

We'll be dealing with the place of status in society and in your life throughout the remainder of this book, but I want to look at one more aspect of the matter before we conclude this present discussion.

How Others Define Us

Thus far, we haven't considered how we get the statuses we occupy, although the answer is obvious in some cases. Your race, gender, and many other statuses were inherited at birth. Sociologists call these **ascribed statuses**. By contrast, **achieved statuses** are those you pick up later in life.

Take a moment to see whether you occupy the status "beautiful person," "average-looking person," or "unattractive person." What about the statuses "optimist" and "pessimist"—

which best describes you? "Success" or "failure"—which status is most appropriate to you?

Now, forget the answers you may have just given. My real question is "Where did those statuses come from?" How do you know whether you are a success or a failure, for example?

George Herbert Mead, an American sociologist prominent at the University of Chicago during the 1920s and 1930s, was particularly interested in this question.[3] As a key founder of the interactionist paradigm in sociology, Mead focused on the power of social interaction in our lives. He said our concept of **self** came from our interactions with others. In particular, he spoke of the peculiarly human ability to "take the role of the other."

Sociologists distinguish between role-*playing* and role-*taking* in this regard. Role-playing refers to those behaviors expected of you because of the statuses you occupy: studying if you are a student, for example. Role-taking, by contrast, occurs when you act out some status you don't actually occupy: children playing cops and robbers, for example. More than a simple diversion, however, role-taking also can offer new perspectives on life.

Looking at the world through another's point of view, Mead felt, is critical for the possibility of society. As we discover how others see things, it becomes possible to reach agreements for living together, indeed to develop shared meanings about life. Yet nothing is more potent than to look at *yourself* through another's eyes. Going one step further, Mead spoke of taking the role of the **generalized other**, gaining an overview of how "everybody" seems to see things: including how they see you.

Just suppose everyone around you seemed to think you were stupid. They said as much and also treated you as though

---

3. See George H. Mead, *Mind, Self, and Society,* Chicago: University of Chicago Press, 1934. The book was edited by Charles W. Morris, a student of Mead's, working from his own lecture notes and those of classmates. Mead, himself, published relatively little.

you were stupid. Suppose the other kids in your elementary
school classes always giggled whenever you spoke in class.
Suppose your teachers never called on you for the answers.
What if you got only *Ds* and *Fs* on your report cards, and your
parents said you should be moᵣᶜ like your brother and sister?
Chances are virtually certain that you would conclude you
were stupid. You'd assume the social status, "stupid person,"
and begin acting accordingly.

A colleague of Mead's, Charles Horton Cooley, spoke of
our **looking-glass self** in this connection, suggesting we see
"who we are" in the reflection provided by those around us.[4] It
doesn't necessarily matter, by the way, if what others "see" is
accurate. More than one young child with defective sight or
hearing has been labeled "retarded," only to take on that status
as a self-image. In earlier times, epileptics were sometimes
seen as "witches," and as a consequence, some took up the
behaviors associated with that status.

Labeling theory is a contemporary name for the sociological view that people often take on the statuses assigned them
by the opinions of those around them. Thus, a young man
widely and unjustly regarded as a thief may begin stealing. A
young woman, scorned unfairly as sexually loose, may eventually loosen up her sexuality to match the image.

W. I. Thomas, a member of sociology's "Chicago School"
along with Mead and Cooley, is most remembered for his
assertion that "if men define situations as real, they are real in
their consequences."[5] All this points to the powerful influence
others have in determining the statuses that shape our experiences of who we are. Ultimately, however, our statuses seem
to conceal rather than reveal who we really are.

"Who am I?" may be a useful question for you to continue
pondering. Don't worry if you can't find an answer. No one

4. Charles Horton Cooley, *Social Organization,* New York: Charles Scribner's,
    1909.

5. William I. Thomas and Dorothy S. Thomas, *The Child in America: Behavior
    Problems and Programs,* New York: Knopf, 1928, p. 572.

else has been able to define human beings satisfactorily. There's a good chance, moreover, that asking the question is more valuable than finding an answer.

Throughout this book and throughout your life, you will find a continuing strain between your experience of who you really are and how you fit into society. Mead recognized this strain and spoke of the **me** as the aspect of the self that reflects the views of others and the **I** as the aspect of the self that responds to those views—either in accord with them or otherwise.

Ultimately, it is important for you to come to grips with both aspects of the self. Never lose sight of the fact that you *are* a part of society, whatever else you may be. You get to participate with others in society by occupying statuses and behaving in accord with the roles associated with them. In fact, it's even fun as long as you can remember that you are *occupying* your statuses, not *being* them.

With all that in mind, let's continue our examination of your social self.

CHAPTER 3

GROUPS

When the Los Angeles Raiders take to the field, people often talk about the individual stars, such as running back Marcus Allen or defensive lineman Howie Long. Or they speak of combinations of players, such as quarterback Jay Schroeder passing to wide receiver Tim Brown. In years past, Raiders fans raved about other individual players, such as the enduring placekicker/quarterback George Blanda, or about other combinations, such as Kenny "the Snake" Stabler passing to wide receiver Fred Biletnikoff.

Still, the Raiders (like other teams) are more than just a collection of individuals and pairs of individuals. The team has its own reality as a unit. When the play begins, each player has a role to play in relation to every other player, and the "silver-and-black attack" is the composite result. Moreover, although individual players come and go, the team lives on.

Social **groups** have a reality, a being, that is *more than just the sum of the individuals comprising them*. Your family of birth was more than an aggregation including you, your parents, and any brothers or sisters; it was also a social unit. Even if a member of your family died, the family itself survived as a unit. Perhaps your graduating class in high school, a church, or a friendship clique also has that quality for you.

Groups and Non-Groups

With rare exceptions, most of our lives are spent in the company of other people. We live with others, eat with them, work with them. We entertain ourselves in the presence of others. We learn together, love together, laugh together. Few of us will even die alone.

Yet, sociologists do not regard all this being together as a matter of being in groups. For example, sociologists distinguish groups from **aggregations**: simple gatherings of people, such as an audience or a crowd. Thus, when New Yorkers gather on a street corner to watch someone climb the side of a building, that gathering is not a group in strict sociological usage.

Similarly, sociologists distinguish **social categories** from what they mean by a group. Thus, left-handed people do not constitute a group, nor do mothers. Exotic dancers are not a group, they are a category. More generally, the people occupying a particular status do not necessarily comprise a group, as sociologists use that term.

Although there are no hard and fast rules about what does or does not constitute a social group, sociologists generally agree on certain characteristics:

◆ *Shared interests:* Members of a group have concerns, goals, or other interests in common. Thus, the members of a Neighborhood Watch group share an interest in protecting the safety of their neighborhood.

◆ *Interaction:* Group members interact or communicate with one another. Typically, the interaction is face-to-face, but sometimes it occurs long-distance.

◆ *Structure:* Even though it may be extremely informal, groups have some degree of structure: some agreed-on set of statuses and agreed-on relationships among them. At the very least, there is likely to be some leadership structure, even an informal one.

◆ *Identity:* This is a critical aspect of groupness. The members must have some experience of belonging to something larger than themselves and larger than a simple relationship with another person. A group is, in some degree, a *union* of individuals, and a class *re*union is a reconstituting of that union.

As I said at the outset, these characteristics are not iron-clad, nor are they precise criteria for distinguishing what is or is not a group. Moreover, the term *group* is used much more loosely than these criteria would suggest, even by sociologists.

Strictly speaking, for example, black Americans would not constitute a group, since there is no way they can all interact with each other, nor do they have a structure. Still, we often speak of blacks as a *minority group* or as an *ethnic group*.

Although it might be better to speak of a minority or ethnic *category* in this instance, the use of *group* is not altogether inappropriate, due to the importance of "black American" as a shared identity, representing a set of shared interests. It becomes all the more appropriate to the degree that blacks have organized themselves (created structures) to pursue those shared interests.

By the same token, it would not have been very useful a few decades ago to speak of homosexuals per se as a social group. Rather, homosexuality was a characteristic shared by certain individuals; it was the basis for a status and a social category. The regular patrons of a particular gay bar might usefully have been regarded as a group, but all homosexuals would not have been regarded as a group. With the recent growth and organiza-tion of "gay rights" concerns and with increasingly common and open identifications with the gay lifestyle, it becomes more and more useful to regard some gays as a group.

Groups and Persistence

Earlier, I pointed out that groups have a reality or "beingness" that is greater than the sum of the individuals comprising them, but that could also be said of, say, an audience or a mob. I'm sure you've been in an audience that somehow seemed to come together as one; we sometimes say the audience "came alive," laughing or cheering together. Performers know this power in audiences well. Perhaps you've found yourself in the midst of an angry mob; maybe you've been swept up in a mob action. I know you've at least seen mobs in action on television. Mobs and audiences definitely constitute a reality of their own. But then, so do relationships. Your relationship with your father, say, or with your best friend exists as something more than just the two of you as individuals.

One of the things that makes groups special is that they *persist,* even though the individuals involved come and go. You'll recall from Chapter 2 that we noted how you have an experience of persisting as *you,* even though the cells of your body, your knowledge, experiences, memories, and everything that describes you changes. Something similar is true of social groups. Consider the example of the very exclusive (that is to say, small) Third Avenue Health Club, shown in Figure 3.1.

Notice that the Third Avenue Health Club, which we began observing in February, is still in existence in June—despite the fact that none of the original members still belong (not to mention that the club has been taken over by nudists). The U.S. Senate illustrates this same phenomenon (persistence, not nudity). The Senate has been in existence for about two centuries, but the current members are completely different from those who were there at the start (with a couple of possible exceptions). The same could be said for the Supreme Court.

This is not to say that all groups last forever; they don't. Still, they have the potential for persistence, a quality groups share with organizations, institutions, and societies, as we'll see later.

Primary and Secondary Groups

Charles Horton Cooley, who, with George Herbert Mead, is credited with the establishment of the interactionist paradigm in sociology, is best known for distinguishing two fundamental

Figure 3.1
The Persistent Third Avenue Health Club

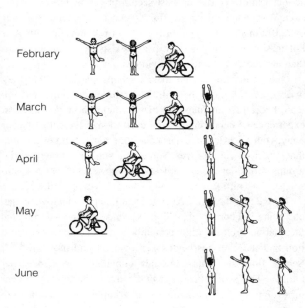

February

March

April

May

June

kinds of groups based on two kinds of relationships. Cooley was particularly interested in what he called **primary relation-ships**, which he described as "those characterized by intimate face-to-face association and cooperation. They are primary in several senses, but chiefly in that they are fundamental in forming the social nature and ideals of the individual."[1]

Primary relationships, then, include those you have with your parents, brothers, and sisters, with a mate, or with your closest friends. Cooley went on to say that **primary groups**, made up of people connected to one another through primary

1.   Charles Horton Cooley, *Social Organization*, New York: Charles Scribner's, 1909, p. 23.

relationships, were characterized by a special "we-feeling." Members of a primary group know each other in something approaching their totality, as whole persons rather than merely as one or two of their statuses.

Secondary relationships, by contrast, are based specifically on the statuses people occupy and the role expectations connecting them. In all likelihood, you have a secondary relationship with the police officer who writes a speeding ticket to acknowledge your creative driving habits. All that matters are your respective statuses of driver and officer. You know nothing of the officer's family life, accomplishments, and aspirations. And if you try to interject information about how important it is for you to be on time for your date because you were late the last time, you are likely to find the officer uninterested and unmoved.

Cooley never actually spoke of **secondary groups**, although they are an obvious contrast to primary groups. Much of your life is spent in secondary groups. If you are a student, each of your classes could qualify as a secondary group. If you work in an office or in some other work group, that may be considered a secondary group. The key quality of secondary groups is that you relate to other group members on the basis of your statuses, against a backdrop of the group's goals and purposes. If you're the receptionist in an office, for example, your duties and performance in that status are more relevant to others in the office than any of the other qualities and characteristics that you may think are more central to who you really are. What matters most is how your status fits into the organization and operation of the group as a whole.

The dividing line between primary and secondary groups and relationships is not a firm one. As in the case of many of the concepts we are going to be examining, there is more value to be gained from seeing the qualities of social life they reveal than from using them to make rigid designations and divisions. Thus, for example, you may be involved in a school class or a work group that is very much a primary group in its intimacy, closeness, and we-feeling. And although it may be a primary group for you, it may not be one for all the other members of the group.

Sociologists have been particularly struck by the apparent need people have for primary groups. Specifically, they have noted the tendency for primary groups to form within secondary groups. Suppose you were one of a hundred employees in the accounting department of a large corporation. That department certainly qualifies as a secondary group, and yet you may become close friends with three or four other employees. You all have lunch together every day, you run errands for each other, and you may get together socially outside of work. That small friendship group would constitute a primary group in the midst of a large secondary group.

An interesting field of study concerns the ways in which such primary groups affect the more formal functioning of the larger group. In some cases, the needs of the two may conflict, as when close friendships serve to protect the job of someone who is not performing effectively (from the corporation's standpoint). Similarly, people sometimes get promotions based more on their primary relationships than on their abilities. At the same time, primary groups can provide alternative lines of communication and novel ways of getting the official work done. If your accounting department needs a new desk computer in a hurry to complete an important job, it may help if someone in the accounting department has a close friend in the purchasing department.

Notice how naturally the interactionist paradigm seems to suit the study of primary groups: the face-to-face interactions of intimately related individuals expressing their joint identity— that we-feeling. As we turn to the more formal, secondary-group structures, there is often more need for the social systems paradigm, focusing attention on the structure and function of relations among anonymous statuses. This is not to say that the interactionist paradigm can't be usefully applied within secondary-group settings, nor that a social systems approach can't be profitably used to understand small, primary groups. Still, there is a natural affinity between the interactionist paradigm and the microsociological analyses, on the one hand, while macrosociological matters seem more likely to call up the social systems paradigm.

The interplay of primary and secondary groups and relationships is a subject of enduring interest for sociologists. Let's look at some other aspects of group life that sociologists examine.

Reference Groups and Identity

As I've suggested already, group membership is an important source of social identity. We've already seen that *who you are* in other people's eyes (and in your own eyes as well, perhaps) is partially a function of the statuses you occupy or the social categories you belong to. It is also a function of the groups you belong to. Thus, if you were on the football team or the cheerleader squad in high school, that membership figured importantly in your high school identity.

Notice there is a definite *circularity* in all types of socially based identity. First, we have certain images or expectations about various social groups. When I mentioned "cheerleader" previously, I know you already had some mental images about what that entailed, regardless of whether those images were positive or negative (or mixed). Second, those *preconceptions* about a particular status or group are *self-fulfilling* in two ways: (a) we tend to perceive people as confirming our expectations regardless of how they actually behave, and (b) those who assume a particular status or join a particular group are likely to do so out of an affinity for the preconceptions involved, and they tend to tailor their behavior to correspond with those preconceptions.

Cheerleaders, for example, are generally regarded as, well, cheerful. That characteristic can be held positively as a "positive outlook on life" or negatively as "bubble-headed." Either way, non-cheerleaders are likely to perceive the cheerleaders they meet as cheerful. Those who become cheerleaders, for their part, are likely to be pretty cheerful to begin with and are likely to shape their future behavior in that mold, both to fit in and to "uphold the honor of cheerleaders everywhere." Please don't regard any of this as peculiar to cheerleaders (positive or negative). It represents one of the fundamental constants of social life.

Sociologists use the term **reference group** to describe groups to which we *refer* in assessing our particular situations. If I am a cheerleader, then, I will determine whether I'm "cheerful enough" by comparing myself to the other cheerleaders, not to people in general. Or as a writer, I've certainly written more books then the average plumber, but it's more relevant to know whether I've written as many as the average writer or the average sociologist (two of my reference groups).

Note in these comments that the meaning of the term *reference group* commonly violates one of the qualities earlier associated with *group* in general. Specifically, many reference groups would appropriately be termed *social categories,* based on statuses. Thus, although I am a writer, I don't spend all that much time with other writers; I don't belong to any writer clubs, and most of my friends are not writers in the normal sense of that term (though many of them can write). Still, "writers" is definitely a reference group for me. Let someone cast aspersions on writers, and I may rise to "our" defense. And when I hear someone speak glowingly about the writer's ability to paint pictures and spark new thoughts with words, I feel good inside, even though the speaker may have Ernest Hemingway in mind.

In this fashion, reference groups become guides for and goals of behavior. Once we've identified with a particular group, there is a tendency thereafter to behave the way people "like that" are expected to behave.

Reference groups can also condition how we feel about our lives. Suppose you were a young professional woman earning

$42,000 a year. How would you feel about that? Before you answer, ask yourself whom you might compare yourself with. For purposes of illustration, let's suppose you compared yourself with other young women in your line of work, and you found your salary was substantially higher than the average. You'd probably feel pretty good about that—you'd feel successful, perhaps, or like someone "on the way up."

Suppose you compared yourself with the young men you knew in your field and discovered most of them were making more than your $42,000. How would you feel then? Sociologists discovered some time ago that a person's relative success or *relative deprivation* was often more significant than their objective circumstances.

In a classic research project conducted in the U.S. Army during World War II, Samuel Stouffer and his colleagues uncovered some truly peculiar findings with regard to soldiers' morale. Based on fairly common stereotypes about the elitism associated with education, the researchers had expected that the more education a draftee had, the more he would resent being drafted. Wouldn't a college graduate, for example, think that being drafted into the Army was really below him? On the other hand, wouldn't the uneducated draftees figure that getting drafted was just their lot in life? The research results were surprising.

In response to a survey that asked whether they felt it was fair for them to have been drafted, those with the most education were actually the most likely to say yes. Those with less education tended to say they should have been deferred—just the opposite of what the researchers had thought. What do you suppose would account for that? Stouffer argued that reference groups and relative deprivation held the answer. Stouffer's line of reasoning was as follows:[2]

---

2. Samuel Stouffer, et al., *The American Soldier*, 3 vols., Princeton, N.J.: Princeton University Press, 1949–1950.

1. During World War II, young men working on farms and on factory assembly lines were often deferred from the draft because their civilian jobs were considered "essential to the war effort."

2. Those young men working on farms and assembly lines were likely to have relatively little formal education. Not many college graduates, for example, would be found in either occupation.

3. Based on 1 and 2 above, we could conclude that young men with little education would be more likely to be deferred from the draft than those with more education.

4. As a general rule, you are likely to pick friends with about the same educational level as you have. Thus the college graduates surveyed were likely to have college graduate friends. Those with only grade school educations were likely to have friends with only grade school educations, and so forth.

5. Based on 3 and 4, soldiers with little education were more likely to have friends who were deferred than would be the case for soldiers with more education.

6. In terms of reference groups and relative deprivation, Stouffer suggested that the soldiers surveyed would decide whether they should have been drafted by comparing themselves with their pre-Army friends. The draftee with little education, and hence with many friends who had been deferred, would probably feel he should have been deferred also. By contrast, the highly educated draftee with few deferred friends would probably feel the draft system was fair; even if he didn't like the Army, he wouldn't feel he had been treated any worse than his friends.

Reference groups provide us with models for who or what we want to become (see Figure 3.2). If you've decided to become a rock musician, then the rock musicians you know, see, or read about provide guidance for your own attitudes and behavior. You may change your hair and wardrobe so that you look like a rock musician. Take a minute to imagine the changes you might undergo if rock musicians were a reference

Figure 3.2
Reference Groups Provide Models for Aspirations

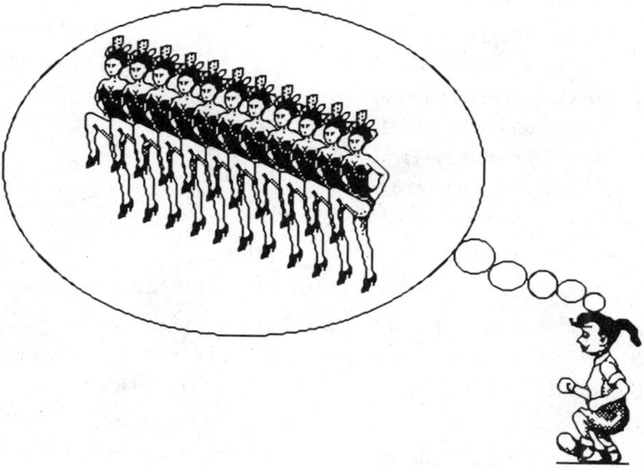

group for you. Then compare that with what you might be like
if you were modeling yourself after Army paratroopers, nurses,
or sociology professors.

In-Groups and Out-Groups

We've seen some of the ways that groups help to define who
we are. At the same time, they define who we *aren't*. If you
take a minute to think about it, you'll see that none of the distinctions we make in describing ourselves could exist without
their opposites. You can't be tall without others being short.
You can't be female without males, and vice versa.

This fact is true of all distinctions. There can't be an "up"
without a "down." You can't be "over here" without there

being an "over there." The fact that distinctions cannot exist without their opposites is also true of our groups.

Rudyard Kipling described the phenomenon in a poem titled "We and They":

Father, Mother, and Me
Sister and Auntie say
All the people like us are We
And every one else is They.
All good people agree,
And all good people say,
All nice people, like Us, are We
And every one else is They.

Sociologists use the term **in-group** in reference to those people who constitute a group or category you belong to and identify with. The term **out-group** refers to the people who don't belong to it. If you happen to have graduated from Harvard University, then Harvard alumni might constitute an in-group for you, with everyone who didn't graduate from Harvard constituting an out-group. Race is an important basis for in-groups and out-groups for some people. Social class is another.

In college towns, members of the campus community and the "locals" often constitute in-groups and out-groups for each other. The same phenomenon occurs in the communities near military bases.

In-groups and out-groups offer more than mere identification. They can determine who interacts with whom. They also provide the basis for intergroup hostilities: prejudice and hatred. Just as "up" needs "down" to exist, it sometimes seems that people can only feel good about the groups they belong to by feeling they are superior to other groups. As Kipling's poem put it: the "good people" and the "nice people" are We. All others are inferior. Therein lie the raw materials for race riots, rumbles between juvenile gangs, wars between nations, bloody religious crusades, and the like.

During World War II, the Nazis attempted to exclude Jews from the human race, and the Japanese took a similar view of the Chinese. And although Caucasian Americans were less brutal to the West Coast Japanese-Americans, the creation of relocation camps in the Midwest offered just as clear a picture of in-group/out-group identities leading to hostilities. Although the wartime governments of Germany, Japan, and the United States might have been excused for imprisoning (perhaps even executing) genuine traitors, they found the scapegoating of racial/ethnic out-groups a more convenient alternative.

The breakup of Eastern Europe during the 1990s has further extended the list of in-group/out-group atrocities. Moreover, the Serbians fighting their neighbors in what was once Yugoslavia have added a new term to the lexicon of hatred: "ethnic cleansing."

Groups and the Definition of Reality

As we've seen, the groups we belong to are central to who other people think we are—as well as to our own sense of identity. I'm sure you were somewhat aware of this already, although you may not have recognized how powerful groups are in that regard. Now we will look at another fundamental influence groups have in our lives. They can shape our very perceptions of what's real and what isn't.

Half a century ago, Muzafer Sherif[3] published the results from a series of experiments that should cause you to question everything you experience and believe. His research addressed the development of group norms in regard to "autokinetic effects."

In Sherif's experiments, a group of people were placed in a totally darkened room. Mounted on a far wall was a small, stationary point of light. I've pointed out that the light was sta-

---

3. Muzafer Sherif, *The Psychology of Social Norms,* New York: Harper & Bros., 1936.

tionary because if you were to sit in a totally darkened room with no visible walls, floor, or ceiling to serve as a frame of reference, you would eventually get the feeling that the point of light was moving around. That's what Sherif's experimental subjects experienced.

Besides signaling the experimenter whenever they saw the light move, the subjects were also asked to estimate the distance it moved. Although the light was stationary throughout the experiments, the estimates of how far it moved varied greatly. More surprising, however, was the discovery that whenever a group of subjects viewed the light together, they quickly reached an agreement about the distance it moved, even though the initial estimates of different members of a given group disagreed greatly. By the end of each observation session, all members of the group were experiencing the same amount of movement. (Different groups, incidentally, arrived at quite different agreed-on estimates of the movement.)

How does this artificial experiment differ from normal social life? The main difference, I suggest, is that we know the group agreements were made up—that the light really wasn't moving. In real life, we make up the same kinds of agreements, and it's difficult sometimes to recognize them as anything other than the truth.

This process is often easiest to see among children, since they aren't as sophisticated as adults in covering their tracks. Maybe you've seen this kind of situation (or perhaps even recall it from your own childhood): a child comes home from school or from an informal gathering of friends and announces that a particular kind of doll or action figure or cartoon character is the best there is. Typically some money must change hands in order to bring the child into harmony with this fact of the universe.

We smile and feel superior when we hear a young child announce with great authority that a particular toy is "cool" whereas another is "dumb." From our vantage point, we can see that the child's opinion is purely a function of what the group has decided. No matter how real it all seems to the child, we can recognize his or her experience as growing out of a process of group agreement.

We are less likely to view matters from that same detached vantage point when the focus shifts to the best/worst automobile, computer, detergent, college, deodorant, religion, form of government, or wine to serve with scampi. You may regard some of these as simply "matters of opinion," but you may not take others that lightly. Your choice of religion, for example, may seem to be based on something that seems much more "real" than mere group opinion—even though you are likely to have the religion you inherited from your parents. (If your choice is "no religion," that will seem *very* real. If you gave up religion, that was probably a part of what you called "being realistic.")

As we saw in Chapter 1, we humans have a tendency to *reify* our opinions—to treat them as though they represented some kind of ultimate truth or reality. And in Chapter 2, we saw how important opinions seem to be in our definitions of who we are: both in the images we present to others and in how we actually feel about ourselves. The key point of this chapter is that the ideas we come to reify and use to define ourselves often originate in groups.

Although we think we perceive and experience life as an individual, personal matter, I want you to see more clearly how the "social construction" of you and those you interact with makes the foundations of your life less personal than you think. And whereas the interactionist perspective is particularly apt in the examination of face-to-face interactions and small group dynamics, the creation of larger and more complex social forms, as we shall see, requires something of a quantum leap. We'll be making increasing use of the social systems and conflict paradigms as we turn now to the study of *organizations* and *institutions*.

CHAPTER 4

ORGANIZATIONS

The Christmas Donation Project was created in 1972 in San Francisco. That year, the employees of a small firm, along with family and friends, spent Christmas Day delivering gifts to patients in San Francisco hospitals, and they joined together for Christmas dinner that evening. The experience was so rewarding that they resolved to make it an annual event. The second year, four hundred volunteers visited two thousand confined children and adults on Christmas Day.

In 1974, the project spread to Los Angeles, New York, and Honolulu, and more people participated. The following year, the project was renamed The Holiday Hospital Project and expanded to include visits during Hanukkah as well as on Christmas. As in earlier years, the volunteers chipped in to buy simple gifts, such as socks, perfume, and small toys. The project continued to spread geographically and to grow in each locale.

In 1980, the project was formally established as The Holiday Project, a nonprofit charitable corporation in California, with local committees in seventy-seven communities across the nation plus affiliated committees outside the United States. That year, 30,000 volunteers visited 242,000 people in 1,144 hospitals, prisons, convalescent homes, detention centers, and similar facilities. Cash donations now totaled $130,000, and another $100,000 was donated in goods and services by businesses. The project was coordinated nationally by a paid executive director and was overseen by a board of directors.

In 1981, four governors and twenty-nine mayors proclaimed a "Holiday Project Day." Some 38,000 volunteers in 114 communities visited 207,000 people in 1,781 facilities, while cash and in-kind contributions exceeded half a million dollars.

Nine years after the project's beginning, the local experience was much the same, but the structure of the organization was very different. The executive director was assisted by a staff of volunteers in the national office. Two part-time national coordinators, operating out of New York City, held weekly conference calls with nine regional coordinators, who were in weekly communication with their local committee chairpeople. Each local committee included a lawyer, an accountant, a media professional, and others responsible for enrollment, fund-raising, institutional contact, and gift-wrapping. This organizational complexity reflected several factors.

First, the participants in the project across the country wanted to see their activities expanded. They wanted more and more shut-ins visited, and they also wanted to share the experience of participating with others. To accomplish this, a national office was needed to make contacts in new cities and to provide assistance and guidance in the organizing of new committees. In addition, a national office—representing tens of thousands of volunteers—had more leverage in attracting media attention and gaining business and governmental support.

Second, a formal national structure could ensure that anything calling itself "The Holiday Project" would be true to the original intention of the project: individuals sharing the experience of the holidays with others. By legally "owning" the name of the project and officially chartering new committees, the national office could prevent it from becoming commercialized—like Santa Claus—and could protect it from fraudulent individuals.

Finally, the project became more formal in response to government regulations. As a legally incorporated charity, The Holiday Project is required to file financial reports with the State of California. How much money was collected, and how was it spent? The purpose of such reports is to protect the general public from fraudulent charities. With tens of thousands of volunteers across the country collecting and contributing, however, adequate record-keeping became a real challenge. The national office had to establish standardized reporting procedures for each local committee, so that all the local records could be combined at year's end to prepare a national accounting.

As with any large voluntary association, perhaps, these organizational requirements established by the national office seem to conflict with the intentions that originally sparked the organization. The first group of volunteers needed no one's permission to begin; they simply did what seemed appropriate. Today, a group of individuals wishing to start a Holiday Project committee would need to comply with the national guidelines and gain official approval from the board of directors. Whereas the first volunteers simply threw some money in a hat and bought presents, each contribution must now be recorded, and receipts must be kept to document all purchases.

Any successful voluntary association will face the problem of an apparent conflict between the interests of the national office and the local chapters. This division is usually clearest in matters of finance. In organizations that are sustained by annual dues from members, one of the most delicate issues becomes how much of the money goes to support the national office and how much goes to support the activities of the local chapters.

Locally, the expenses include telephones, postage, office supplies, and perhaps office rent. Nationally, the expenses are those of any formal organization, including staff salaries. In the case of a charitable organization such as The Holiday Project, there are numerous "hidden" costs associated with the legal issues mentioned. For example, the filing of legal papers often involves the cost of attorneys and filing fees. The need for careful financial accounting includes annual audits by accounting firms. Liability insurance is required to protect against any injuries that may occur in the course of the visitations. Coordinating all these and similar administrative activities entails additional costs for telephoning, mail, and perhaps travel.

The various expenses needed to run a national organization can seem pretty far removed from the original purpose—to visit people in hospitals and other facilities during the holidays, in the case of The Holiday Project—and yet the local committees are expected to raise the money to pay those expenses.

Quite aside from the question of who gets the money, there are other sources of local/national conflicts. The legal requirement for financial accounting results in local committees having to prepare and file numerous reports. Whereas the participants in the original group in 1972 could simply chip in a few dollars each and spend it however they saw fit, all contributions and expenditures must now be reported to the national office on a monthly basis. Moreover, the national office must exercise some control over the acceptability of expenditures and even fund-raising techniques by local committees.

The Holiday Project has continued to avoid the establishment of a large, paid, professional staff, choosing instead to operate as a gathering of volunteers. During 1991–1992, a total of 15,442 visitors in thirty states visited 135,007 people in 1,423 different institutions.[1]

There are other dimensions to the local/national division. In the case of The Holiday Project, the "work" of the organization—people visiting hospitals and other facilities—must ultimately be done at the local level. In other organizations,

---

1. The Holiday Project, "1992 National Conference Notes," June 24–28, 1992, p. 1.

however, this is not so clear. Consider an organization such as Zero Population Growth (ZPG), dedicated to dealing with the problem of overpopulation. Much of ZPG's work has been done by local chapters engaging in public education at the local level. At the same time, much of the work—on national policies such as immigration, abortion, and tax exemptions for dependents—is more appropriately addressed at the national level. In such situations, conflicts may arise over where the greatest effort (and resources) should be directed. Again, the local chapters are often the focus of fund-raising through new members' dues, and those local chapters can feel "ripped off" when they have to send a substantial portion of those funds to the national office.

All I've said about voluntary organizations is paralleled in large for-profit corporations as well. There, the conflicts can occur between headquarters and the branch offices, as well as between organizational divisions such as personnel, marketing, and production. Such problems, in short, are inevitable in the operation of organizations.

One element in the shift from people simply cooperating in the interest of a common goal to participating in a formal organization can be seen in how they refer to each other. Whereas the original participants knew each other as Sam, Pat, or Mary, participants now often refer to the statuses they occupy. In the case of The Holiday Project, participants in the local committees are likely to speak of "sending money to national" or of deadlines set by "the board of directors." Those on the board of directors or on the national committee, for their part, are likely to refer to "the locals" or "Boston." As the organization has grown more formal, it has become more impersonal in this sense.

As we'll see in this chapter, an organization is a powerful tool for humans to get things done. We'll also see that organizations tend to take on a reality of their own, sometimes overshadowing the humans involved and even supplanting their original purposes. You should also see how a sociological view of organizations can make you a more critical observer and can help you understand why organizations sometimes operate in the odd ways they do.

What's an Organization?

Although all of us talk with ease about organizations and participate within many, it's not all that easy to define exactly what an organization *is*. It may be useful, therefore, to begin with some of the things an organization *is not*.

Let's consider an organization such as International Business Machines (IBM), a giant industrial corporation employing hundreds of thousands of people engaged in a variety of activities. It's clearly an organization, but what aspect of IBM is "the organization"? To begin, IBM, the organization, is more than its name. In fact, it began as the Tabulating Machine Company.

Similarly, IBM is not the buildings it occupies nor the products it manufactures. And although it may be seen legally as its papers of incorporation, the organization is clearly far more than that.

It's tempting to say that IBM is its employees, but that's not exactly accurate either. From a sociological standpoint, it would be closer to the truth to say IBM is a table of organization, a structure of statuses and the specification of role relations among them. Said somewhat differently, IBM, the organization, is a set of agreements (rules) that form a context within which people interact to achieve certain objectives. It is a social map describing how a particular collection of people are to work together.

There is another possibility for people coming together for a common purpose. I'm sure you've experienced it, in fact. Perhaps you and some friends decided to throw a small, spontaneous party. One of you may have gone shopping for refreshments, another cleaned up, and someone else invited people. Or maybe you all did a little of each chore. Overall, each individual just did whatever was needed for the group, without the need for anyone to be in charge. Maybe you've experienced something similar in a different situation. It often happens in sports and also in disaster situations. That kind of spontaneous, unorganized cooperation is sometimes called **alignment**: a group of individuals sharing a common purpose with each doing whatever is appropriate to achieve it.

Alignment doesn't seem to work very well when the number of people involved gets large or when the work to be done is complicated. It's hard to conceive of all the IBM employees coming to work and simply "chipping in" to build computers. Instead, the many specific functions required to achieve the whole purpose are specified and named, with individuals learning to perform those specific functions. And leading, managing, supervising, and organizing are functions that need to be specified and performed.

Organizations, like relationships and groups, provide another aspect of social identity for individuals. The organizations we participate in become another way of representing who we are. At the same time, we create identity at a collective level, establishing entities that become actors within society. These new collective entities may supply us with some of the things we want in life, and they may also make demands on us: organizations may require us to "pay our dues" both literally and figuratively.

Bureaucracy

The epitome of formal organization is **bureaucracy.** It is a familiar term for most Americans, who commonly distinguish among several types of bureaucracy: "stupid bureaucracy," "petty bureaucracy," "bungling bureaucracy," "damned bureau-cracy," and worse. For most people, bureaucracy means long lines, red tape, impersonality, forms, and frustration, although bureaucracies don't have to have all these characteristics.

Max Weber (1846–1920) did the earliest comprehensive analysis[2] of this form of organization, and his observations are still generally applicable to today's bureaucracies. He saw the following as the key characteristics of bureaucratic organization:

1. Jurisdictional areas are officially fixed and generally gov-erned by the rules. The required regular activities of the organization are specified as official duties. The structure of authority in the organization is clearly spelled out, as are the means by which each member of the organization is to fulfill his or her official duties.

2. The organization is structured hierarchically, with specified levels of authority. Like a pyramid, the organization has fewer people at each higher level of authority, ending with a single person at its head.

3. The organization is managed largely through the prepara-tion, maintenance, and use of written files.

4. Those holding positions in a bureaucracy are specially edu-cated for their jobs.

5. The fully developed bureaucracy requires the full working capacities of those holding positions in it. Administration is a full-time job. In this regard, Weber contrasted modern bureaucracies with earlier enterprises in which management was treated as a secondary activity.

6. All of the organization's official activities are governed by formalized rules, and knowledge of the rules constitutes a special skill.

In summary, bureaucracies engage in the business of "administration," and they do so through officially specified

---

2. Hans Gerth and C. Wright Mills (eds.), *From Max Weber: Essays in Sociology,* New York: Oxford University Press, 1946.

agreements regarding status relationships and role expectations, which have little or nothing to do with the specific humans who occupy the statuses. Alfred Krupp, Hitler's chief munitions maker during World War II, put the matter more bluntly:

> **What I shall attempt to bring about is that nothing of importance shall be dependent upon the life or existence of any particular person; that nothing of importance shall happen . . . without the foreknowledge and approval of the management; that the past and the determinable future of the establishment can be learned in the files of the management without asking a question of any mortal.[3]**

Although it is a simple matter to find fault with the cool impersonality of bureaucracy so described, it is important also to recognize its many advantages. Weber felt bureaucracy to be technically superior to other forms of organization in the same way that machine production is technically superior to handwork. Bureaucracies are faster, more precise, clearer, more certain, more effective, and more efficient than nonbureaucratic organizations.

In addition, Weber saw an advantage in the continuity of bureaucracies and their files. Individual members of a bureaucracy can come and go, but the structure of statuses and written records of past activities give the organization a life of its own that outlives the service of individual bureaucrats. The continued functioning of the U. S. civil service in spite of the frequent changing of politically appointed officials is an illustration of this feature of bureaucracies.

Weber cited the technical expertise of members of a bureaucracy as another advantage of this form of organization. Only in a bureaucracy is it possible for people to develop extreme competence in very narrow specialties and for all those special skills to be coordinated into a productive organizational effort.

---

3. Quoted in Alvin Gouldner, *Patterns of Industrial Bureaucracy,* Glencoe, Ill: The Free Press, 1954, pp. 179–180.

Bureaucracy Run Amok

Although it's important to recognize the advantages of bureaucratic organization, we shouldn't overlook its foibles. Let's
look at some of the problems people have pointed to.

In 1957, C. Northcote Parkinson drew attention to what he
modestly called "Parkinson's Law."[4] Parkinson stated his law
as follows: "Work expands so as to fill the time available for
its completion." No matter how little work is to be done or
how much time is available, the work will be stretched out
enough to fit perfectly within the time at hand.

Bureaucracies are characterized by regular work hours and
by specified activities to take place during those hours. If too
much work is expected, this discrepancy will be made known,
and either the work will be reduced or the staff will be
increased to handle it. If, on the other hand, there is not enough
work to be done by the staff assigned to it, that work somehow
keeps everyone "busy" throughout the working hours.

Parkinson's Law appears to extend to the case where there
is no work to be accomplished. Sometimes a change in the
structure of an organization will leave individuals with no useful function in the organization. Typically, it is possible to
stretch "nothing" enough to fill the day. Indeed, whole organizations may find themselves with no useful function.
Bureaucratic organizations seldom die, however. Every year
the government creates many new agencies, usually with man-
dates to solve specific problems. Rarely, if ever, does such an
agency solve its problem and ask to be dissolved.

In *The Peter Principle*,[5] Laurence Peter stated his "principle" as follows: "In a hierarchy every employee tends to rise to
his level of incompetence." If promotions in a bureaucracy are
based on demonstrated merit, a person who proves capable of
doing a particular job with excellence will be removed from

4. C. Northcote Parkinson, *Parkinson's Law,* Boston: Houghton Mifflin, 1957.
5. Laurence Peter and Raymond Hull, *The Peter Principle,* New York: Morrow, 1969.

that job and elevated to one in which he or she has not demonstrated excellence. Often, of course, the person quickly learns
the skills required in the new position and again performs with
excellence—and is promoted once more. Peter suggested that
members of a bureaucracy will continue to be promoted until
they reach positions in which they perform terribly. Then they
will be left in those positions, because further promotions are
clearly not warranted. Many bureaucrats, then, will occupy
positions that outstrip their competence, since demotions are
rare in bureaucracies.

Criticisms of bureaucracy are often expressed in terms of
what's wrong with "the system." American radicals sometimes complain more specifically about "the capitalist system," while Russians standing in long lines for toilet paper or
shoes used to mumble (quietly) about "the socialist system."
(Now some of the same people blame their troubles on the
shift to capitalism.)

John Gall, in his 1986 *Systemantics,* suggested the problem
lay with *systems* per se.[6] Despite the undeniable advantages of
being systematic in organizing certain processes—building
cars or computers, for example—Gall suggested that the systems we create tend to turn on us. Here's an example:

> **The largest building in the world, the Space Vehicle
> Preparation Shed at Cape Canaveral, *generates its own weath-
> er, including clouds and rains.* Designed to protect space rock-
> ets from the elements, it pelts them with storms of its own.[7]**

Human systems, moreover, are no less perverse than physical ones. A big part of the problem relates to organizational
survival. Gall stated three related axioms:

♦ Systems Develop Goals of Their Own the Instant They
   Come into Being.
♦ Intrasystem Goals Come First.

---

6. John Gall, *Systemantics: How Systems Work and Especially How They
    Fail,* Ann Arbor, MI: The General Systemantics Press, 1986.

7. Gall, p. 24. *Emphasis in the original.*

◆ The System Behaves as If It Has a Will to Live.[8]

You may recall from Chapter 3 how groups are able to survive well beyond the lifetimes of any members who pass through them. (Recall the Third Avenue Health Club, for example.) Formal organizations possess this capacity more than simple groups. And although Gall is speaking somewhat tongue-in-cheek about a system's will to live, what he says is essentially true. For although "will" is a psychological quality, it is nonetheless possible to structure organizations so they behave as though they had a will to survive. In fact, organizations that were not structured that way aren't around for us to study today.

It could be said that the first purpose of any organization is its own survival. It even makes sense to have it that way.

Suppose you and I are committed to feeding the homeless people in our hometown. The two of us don't have either the money or the time to get the whole job done on our own, so we create a local organization called Feed the Homeless. We ask people for contributions to help buy the food, and we get people to help us deliver the food.

I think you can see how an organization like the one I've described could accomplish something worthwhile that you and I could not accomplish as two individuals. At the same time, it's too obvious to mention (which doesn't stop my doing so) that if the organization ceased to exist, it would not accomplish its purpose.

Now enters the government. We are informed that if we are going to solicit contributions from people, we will need to account for where the money comes from and how it's spent. Notice that this is a perfectly reasonable requirement, since it protects you and me from fraudulent fund-raising activities.

Thus, for our organization to continue, we must now devote some of our efforts to the maintenance of proper financial

---

8. Gall, p. 74.

records. To be sure the job gets done correctly, in fact, we may hire a part-time bookkeeper to keep our records. However, this means that some of the money we raise must go to support this activity.

Now let's suppose we are given some food that has gone bad: we distribute it to the homeless and some of them get sick. Fortunately, no one dies, and the sick people are persuaded not to sue us. It becomes clear, however, that our whole organization could be wiped out by that kind of problem. So we decide we must buy some liability insurance. Again, we are taking an action necessary to keep the organization in existence so it can do its work.

All this makes perfectly good sense, and it illustrates how organizations' goals take priority over the original purposes, thereby somewhat frustrating our efforts to accomplish what we initially set out to do. Moreover, this same dynamic process creates strains between the organization and the individuals who inhabit it.

Organizations and Individuals

It is often argued that modern organizations damage the individuals who comprise them. Nearly four decades ago, William H. Whyte wrote a popular sociological book titled *The Organization Man,*[9] in which he examined the many ways in which modern corporate employees were forced to give up their values, their families, and their individuality to succeed in the corporate world. They had to dress alike, drink alike, talk alike, and think alike in order to be members of the team.

9.   William H. Whyte, *The Organization Man,* Garden City, NY: Doubleday, 1956.

10. Gareth Morgan, *Images of Organization,* Newbury Park, CA: Sage, 1986, Chapter 7.

More recently, Gareth Morgan has written that sometimes organizations can be usefully seen as "psychic prisons."[10] Some organizations operate as repressive, paternalistic families, for example. Such organizations may satisfy the psychological needs of some participants, while creating problems for others.

Much attention has also been given to the notion of "burnout," often attributed to conflicts between the demands of organizational and personal lives. This occurs both in the organizations we work for and in the voluntary associations we serve on an altruistic basis. Too often, it seems as though our organizations call for more than we can deliver on a sustained basis. The demands for time and emotional energies often leave us drained and unable to enjoy life in general.

In part, this conflict is simply inevitable. Our participation in organizations takes time that we might have wanted to spend elsewhere. At the same time, we may not like the work we have to do in an organization, but the organization cannot survive without someone doing that work. Someone has to take out the trash, file the tax reports, fire the problem workers. And whereas specialization is an efficient way to organize, some people spend all their time doing those things no one wants to do.

This problem is all the more troublesome when the "housekeeping" chores seem far removed from the main purpose of the organization. Margaret Sanger, the woman largely responsible for the beginnings of birth control in America, used to say she had no trouble getting volunteers to hand out pamphlets or do public speaking—her problem was getting someone to sweep out the office.

If all this weren't bad enough, there is an even subtler aspect to the conflict between organizations and individuals. Whereas we have a tendency sometimes to personify "the organization" and see it as somewhat evil, it could be argued that we aren't "good" enough for our organizations. Fundamentally, an organization is based on the assumption that the individual participants will be willing to sacrifice themselves for the good of the organization—and that all will benefit from the organization's accomplishments.

This can be seen clearly in the origins of consumer cooperatives in the United States and elsewhere. The initial idea was that a group of consumers would join together to purchase goods at volume discounts. For simplicity, let's assume that four of us wanted to buy corn on the cob. By buying a bushel from a wholesaler, we could get a lower price per ear than at the supermarket. However, none of us wanted a whole bushel. So we would chip in on the bushel and divide it among us.

The simple elegance of this arrangement would be slightly hampered by the need for someone to collect the money, someone to pick up the bushel of corn from the wholesaler, and someone to distribute it among the four participants. In a single transaction like this, however, we could undoubtedly work out a division of labor informally; small groups of people do this sort of thing all the time.

To do all this on a larger scale—more participants and larger purchases—the amount of administrative work increases as well. And whereas a small group of friends may be able to trust each other to do their fair share of the work, that becomes less likely in a larger group. Soon, some will be doing all the work while the others are getting a free ride. Eventually, formal agreements would be needed to be sure everyone was contributing equally to the whole, and some people would have to be responsible for enforcing the agreements. All this is necessary because we don't feel we can trust each other to be fair— and perhaps with good reason.

Ultimately, we find ourselves suffering through organizations that assume none of us will make a fair contribution unless we are *forced* to do so. And given our distaste for the oppressive character of such organizations, we look for ways to "beat the system." Because some workers call in sick when they aren't, for example, we restructure the rules of the organization to limit each worker to, say, two weeks of paid sick leave each year. Before long, most employees are being "sick" exactly two weeks each year—since they now feel they have it coming.

It's worth noting that bees and ants don't have these kinds of problems. They show up for work every day—holidays included—regardless of the weather or what they did the night before. Each of the individuals does his or her job, each contributing to the whole. If they need to die for the sake of the organization, they do it without complaint.

Bee and ant corporations work more effectively because all the individuals are *aligned* by instinct, a quality that humans do not seem to share with them. This is not necessarily bad, of course. Although the instinctive alignment of bees and ants ensures stability and regularity, it comes at the cost of creativity and innovation. This is a trade-off we'll come back to in later chapters when we consider freedom, order, deviance, and social change.

Humans, then, do not appear to be aligned by instinct. Although humans are sometimes able to align temporarily in small groups, larger organizations require the institution and enforcement of formal agreements.

In the past few decades, a great deal of attention has been given to alternative ways of structuring and managing organizations to avoid the problems we've been examining. In 1960, Douglas McGregor[11] introduced the notion of *Theory Y* management, based on a greater respect for workers as human beings. This notion was expanded in 1981 by William Ouchi's *Theory Z*,[12] which suggested ways American businesses could emulate the successful management techniques of Japanese companies.

In 1985, John Naisbitt and Patricia Aburdene introduced *Re-inventing the Corporation,*[13] in which they spoke of the need for leaders to create a shared vision among employees.

Creating the vision is the leader's first role. Next, she or he must attract people who can help realize it by adopting the

11. Douglas McGregor, *The Human Side of Management,* New York: McGraw-Hill, 1960.

12. William Ouchi, *Theory Z: How American Business Can Meet the Japanese Challenge,* New York: Avon Books, 1981.

13. John Naisbitt and Patricia Aburdene, *Re-inventing the Corporation,* New York: Warner Books, 1985.

vision as their own and sharing responsibility for achieving it. The name of this critical process is *alignment.* In its highest form, alignment creates a remarkable experience, whether it happens on the playing field, in the concert hall, or at work.[14]

When you identify with your company's purpose, when you experience ownership in a shared vision, you find yourself doing your life's work instead of just doing time.[15]

More recently, Peter M. Senge's *The Fifth Discipline*[16] focuses on what he calls the "learning organization." Building from the Greek concept of *metanoia,* Senge looks for ways in which organizations can achieve a "shift of mind" akin to the experiences of transcendence or transformation sometimes experienced by individuals. The benefits of such a shift accrue to both the organization and to the individuals participating in it.

The work of these authors and others suggests that it is possible to create organizations that nurture individuals rather than damage them. At the same time, as we've seen, it does not happen automatically. The task of finding ways to create nurturing and effective organizations is a continuing one in sociology. This task is complicated, however, by the power that **institutions** wield over organizations—as we'll see next.

---

14. Naisbitt and Aburdene, p. 24.

15. Naisbitt and Aburdene, p. 26.

16. Peter M. Senge, *The Fifth Discipline,* New York: Doubleday Currency, 1990.

CHAPTER 5

INSTITUTIONS

In 1965, a young lawyer in Washington, D.C., published a book on auto safety that was to have more far-reaching impact than could have been anticipated at the time. Ralph Nader's *Unsafe at Any Speed* was written with the overt purpose of drawing attention to certain design flaws in Chevrolet's Corvair and other automobiles. Nader asserted that the automobile industry had an obligation to exercise greater concern for automobile safety and urged government action to ensure compliance with that obligation. The main details of what followed are commonly known: the automakers responded with general resistance to the changes demanded and with personal attacks on Nader. Subsequently, numerous auto safety laws were enacted, autos were redesigned, and a court awarded Nader a substantial sum in recompense for the personal attacks.

Although we should not downplay the significance of the changes brought about in auto safety, something more important also happened as a consequence of Nader's challenge of the giant automakers. The climate within which big businesses operate in America was substantially altered. Specifically, Nader brought into existence the possibility that consumers could have a say in the nature of the products they consumed, analogous to the participation labor unions had gained for workers thirty years earlier.

Evidence of Nader's impact is manifold. For example, in 1982, a young Rochester, New York, mother named Leslie

Hughes took her children to McDonald's to pick up something
for dinner. The fast-food chain had just introduced their "Happy
Meals," containing both food and small action figures: a sheriff
with his rifle, an Indian with a spear, and so forth. Leslie
Hughes ordered the Happy Meals for her children's dinner.

After dinner, Hughes examined the small toys and was hor-
rified to discover that the rifles and spears were so small that
her young daughter could have easily gotten them stuck in her
ear, nose, or throat. In addition to taking the toys away from
her children, Hughes went further and called the Empire State
Consumer Association and discussed the matter with Judy
Braiman-Lipson, the group's president.

Once Braiman-Lipson had checked out the toys for herself,
she contacted McDonald's, and the company agreed to set up a
conference for her to meet with the firm that had tested the
safety of the toys for McDonald's. Braiman-Lipson also contacted the federal government's Consumer Product Safety
Commission and alerted them to the problem. The government
began its own tests of the Happy Meals toys.

Exactly one week after Leslie Hughes had purchased the
meals for her children, McDonald's executives were meeting
to decide what to do about the problem. That night and the
next day they reviewed the conclusions reached by the various
private and public agencies now involved. By early evening
telephone calls were being made to all McDonald's franchises
across the country, and all the Happy Meals were withdrawn
by midnight.

I suggest it would have been extremely unlikely for Leslie
Hughes, a Rochester homemaker, to have had anything
approaching that impact without the successful precedent
Nader and others established in integrating "consumer rights"
into America's business climate. The Leslie Hughes story is
scarcely unique today.

Looking beyond business, Nader's action has also brought
about a shift in the nature of "citizen rights," with individuals
challenging government actions in regard to health, environment, and other aspects of life. New laws have been passed as

a consequence, and government officials have been pressured to enforce existing laws.

Although Nader may have set out to bring about changes in the actions of specific businesses, his more important impact has been in shifting the fundamental agreements about the rights of consumers and the obligations of businesses in general. Rather than challenging the actions of a particular government official or even a particular level of government, he brought about a change in the relationship of citizens to government.

As we'll see in this chapter, Nader's most important impact was within the domain of **institutions**—business and government—rather than in the domain of organizations. Institutions, as we'll see, constitute the context within which organizations function, so that changes in institutions affect countless organizations. This is analogous to changing the requirements for graduation as opposed to making an exception for one student, and it points to a powerful leverage point for social change.

Institutional Imperatives

To understand what institutions are, it is useful to begin with a look at why they exist. To start, let's recognize that individuals have numerous needs to be satisfied within society. There are, for example, the requirements of survival: food, water, shelter. Looking beyond mere survival, human beings appear to want to survive with some degree of comfort and to enjoy some feeling of security that they will continue to survive comfortably in the future. Even more intangibly, they appear to want meaningful relationships with others, satisfaction, and other qualities difficult to define yet compelling in their motivations to action. However a society is structured, then, it must deal in some fashion with those things individual members want and need from life.

At the same time, societies—as systems—have require-ments of their own. Although I don't want to personify society by suggesting it has wants and desires, hopes and fears, it is essential to recognize that a society cannot survive unless cer-tain conditions are satisfied. In turning to the needs of society, realize that we have moved fully into the domain of macroso-ciology, as promised earlier.

For example, no society can survive unless it handles the problem of *replacement*. As individual members die, new ones must be born and reared to replace them. Similarly, organized society must make some provisions for leadership, for govern-ment. The list goes on.

Many individual and societal needs are related to one another. In the case of society's need for replacement, for example, the individual's desire for sexual expression and the desire for meaningful relationships are directly relevant. Notice, by the way, that an individual can survive without sex (you really can), but a society cannot. This illustrates that whereas society can be seen as an entity with needs, its needs are not the same ones human beings have. Thus, whatever you may know about human beings is not sufficient for an under-standing of society.

In the best of all possible worlds, structures for living together would be devised that satisfy both individual and societal needs. In the most obvious case, perhaps, the individ-ual's need for sexual expression and for meaningful relation-ships could be satisfied in such a fashion as to meet the society's need for replacement.

As it turns out, the best of all possible worlds is not necessar-ily a pipe dream. Actually, there are numerous structures that can match up individual and societal needs. In the case of sexu-ality, relationships, and replacement, for example, numerous forms of the **family** have been created for that purpose (see Figure 5.1). The form we know best is the monogamous, **nuclear family**, consisting of mother, father, and children. In other societies, such as traditional China and Japan, an **extended family** combines grandparents, parents, and children in one unit.

Figure 5.1
Various Marriage and Family Forms

Nuclear Family

Extended Family

Monogamy

Polyandry

Polygyny

Group Marriage

In some **polygynous** societies, one husband has several wives, and in **polyandrous** societies, one wife has several husbands. Other societies have practiced **group marriage**: several husbands and wives living together as mates to each other. The point is that all such forms have accomplished what was required by the individuals involved and by the society as a whole. The same, moreover, could be said of different forms of government, such as monarchies, aristocracies, democracies, and dictatorships.

Although many forms for handling the needs of individuals and society are possible, it is usually the case that a particular form works only if everyone in the society abides by that one form. For example, it's hard to imagine a stable society in which some people operate under the divine right of kings while the rest are democratically structured. Or in economic terms, either socialism or capitalism seems to work, but a given society must select only one (even if it's a hybrid "mixed economy").

The need for agreement on an institutional form applies in most cases. Even though it would seem that monogamy, polygyny, and polyandry could coexist in a given society, people seem to need to believe their form is best, especially to the extent that they have linked their personal identities to the social pattern. Disagreement over which pattern is best often results in a struggle for supremacy among different forms. Moreover, as we'll see shortly, the various institutions of a society (family and religion, for example) tend to link supportively with one another, and the need for agreement is heightened.

In short, although many ways of satisfying the needs of society and of individuals are possible, there is a powerful tendency to *institutionalize* one form among the many possibilities. This means that one way of doing things becomes accepted, legitimate, customary, expected, even required.

Sociologists traditionally have spoken of five major institutions governing different domains of social life: *family, religion, government, economy,* and *education.* At the same time, it is sometimes of value to apply the term *institution* more specifi-

Figure 5.2
Institutions Are the Context for Organizations

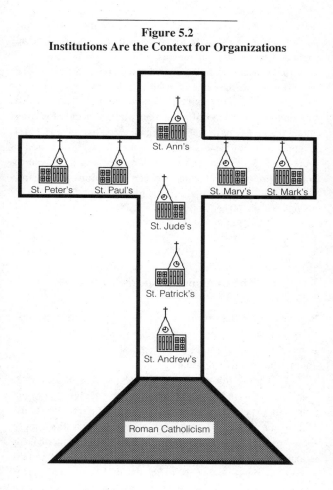

cally. It may be useful, for example, to see democracy, capitalism, the military, labor, higher education, or Christianity as
institutions. This is appropriate whenever we are talking about
a set of organizing principles and not specific organizations.

As Figure 5.2 illustrates, it might be useful to view Roman
Catholicism as an institution: a set of beliefs, values, and prac-

tices. As such, it would form the context for the various individual churches (organizations) that make up the Roman
Catholic Church as a concrete body.

Strictly speaking, then, Harvard is not an institution; it is an
organization. At the same time, there might be some value in
seeing it as an institution if our attention were on "the Harvard
tradition" or "the kind of education Harvard stands for" and
not on the people or the table of organization that give Harvard
its concrete reality as a university you can drive to, attend, or
protest against. For the most part, however, you'll do well to
draw a more rigid distinction between institutions and organizations. This is a more open-and-shut distinction than many of
the others we've discussed earlier.

The Components of Institutions

So far, I've alluded to institutions as a context for organizations, as sets of organizing principles, and as customary patterns. Now I'll get more specific, pointing to four important
components of an institution.

Let's begin with **norms**: expected patterns of behavior.
You'll recall from Chapter 2 that the term *role* referred to the
behaviors expected of someone occupying a particular status.
Norms are almost the same as roles, except that they apply
more broadly across statuses. *Voting* is a norm in America, for
example. In some instances—such as the selection of certain
public officials—voting is officially mandated, with great
specification as to the procedures by which it will occur. We
might even speak of voting as a role associated with the status
"citizen." More generally, however, voting is a widely accepted method for making group decisions in America. If your
bowling club has to decide whether to have Bambi or Thumper
as its team symbol, you'll probably have the members vote.

That procedure would not be so obvious a method in some other societies.

Norms, then, represent the ways we expect people to behave. But how can we be sure they will behave as expected? This is accomplished through the use of positive and negative **sanctions**: rewards and punishments. Like norms themselves, some sanctions are formal and official; others are more informal. Some are harsh, others are gentle. If you get caught tampering with the ballots at the bowling club, you may have to buy the team a round of beers. If you tamper with a federal election, you may not see beer for a long time.

The list of possible examples of norms and sanctions is virtually endless; it's the sea you swim in every day of your social life. To begin, all laws qualify as norms: speed limits, drinking age, taxes, and the prohibitions against murder, burglary, embezzlement, rape, treason, nudity, drug dealing, and so forth. There are federal laws, state laws, county laws, and local laws. Some laws are established through direct elections, others are voted into existence by our representatives, and still others are established by agencies such as the Internal Revenue Service.

Numerous though they are, however, the number of laws pales in comparison with the number of informal norms that govern you: what language to speak, what kinds of clothes to wear, the kinds of foods to eat, what length to wear your hair, what music to like, how loud to speak on the telephone, how much to tip in restaurants, which beverages to mix and which not to, what animals are appropriate as pets, how long to wait before deciding that your date isn't going to show up, and so on. If you don't think these really qualify as norms, you might try leaving as soon as your date is ten seconds late, showing everyone your pet slug, or mixing up a batch of tomato juice and coffee for your friends. Try these also if you don't think there are any sanctions associated with the norms. Just don't say you got the idea from me.

Where do you suppose norms come from? Some seem to have come about purely by chance. A story is told that when

George Frideric Handel's *Messiah* was first performed for the King of England, the king unexpectedly rose to his feet as the rousing "Hallelujah Chorus" began. In England at that time, when the king stood, everyone stood, so the entire concert audience rose and remained standing throughout the chorus. To this day, audiences still rise for the "Hallelujah Chorus," even if they've never heard the story of the origin of the practice. Moreover, nobody knows why the king stood up in the first place. Maybe he thought the concert was over. Maybe he was trying to slip out to the bathroom. Whatever his motivation, a norm was born.

Most norms, perhaps, have a more logical place within society. Specifically, norms are often justified or legitimated on the basis of **values**. In America, for example, our norm of voting is justified by the value we place on broad participation in public decision making. Values are a matter of preference; they represent views about what's better than what, and in America we feel it's better for everyone to participate in decision making than to let the one or the few dictate to everyone else.

Speed limits, as norms, are clearly based on a value of public safety. The norms requiring a certain number of years of schooling are justified on the basis of our valuing an educated citizenry. (Notice, by contrast, that a dictator might prefer to keep his or her subjects ignorant.)

Finally, values are, themselves, justified on the basis of **beliefs**: views about what is true. Our valuing of mass participation, then, is based on a belief that all are "created equal" and "endowed by their Creator with certain inalienable rights." As you'll recall, the original belief was that all *men* were created equal: *white* men, to be specific.

It's important to recognize the relationships among beliefs, values, and norms. As Figure 5.3 indicates, those relationships can be seen in terms of *justification* or in terms of *specification*. As I've already pointed out, values justify norms and beliefs justify values. Moving in the other direction, however, values can be seen to specify what ought to be important if the belief in question is true, and given that value, the norm speci-

Figure 5.3
Components of Institutions

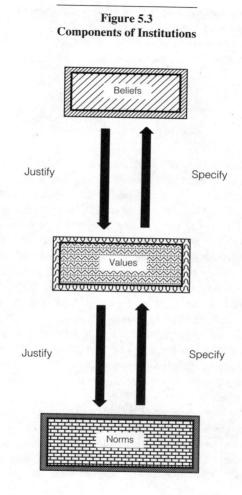

fies appropriate action. Given the belief that all are created equal by their creator, it stands to reason that all should have a say in government. And, given the value that everyone should have a say, voting is a way of putting that into practice.

An institution, then, is a set of beliefs, values, and norms governing some area of social life. Some of these elements relate to one another as already described; other elements may not seem that directly related to others.

Institutional Conservatism

As we have seen so far, institutions are created as ways of satisfying the needs of individuals and also of society as a system. Neither of these is the first task of an institution, however. Whatever individual or societal needs an institution addresses, its most deliberate and unfailing first purpose is to *perpetuate itself.* The first task of democracy, for example, is to perpetuate democracy; monogamous, nuclear families are designed so as to perpetuate the monogamous, nuclear family as an institution. Christianity operates first and foremost to perpetuate Christianity. Hinduism is designed to perpetuate Hinduism.

I don't intend these comments as negative or cynical. This is simply the way institutions are designed, just as your car is designed to go straight if you take your hands off the wheel. Institutions are designed to ensure their own survival. Notice how the linking of beliefs, values, norms, and sanctions serves that purpose.

The obvious function of sanctions, of course, is to perpetuate the norms they are associated with. Sending people to prison for tampering with the ballot box, for example, helps perpetuate the norm that everyone will have one vote, and all votes will count equally. The value of mass participation in public decision making also helps to perpetuate that norm, but notice that the norm also helps to perpetuate the value. A similar pattern emerges when we add beliefs to the mix. To varying degrees, then, institutions are closed systems in which the component parts reinforce one another with a circularity that perpetuates the whole system in the process.

To the degree that an institution is in fact a closed system, it is impossible to argue for change within the constraints of that system. If people accept the belief that God chose the king to rule over them, for example, they are unlikely to demand the establishment of democratic practices. At the very least, they could not base an argument for democracy on a belief in the divine right of kings. Garrett Hardin, a biologist, has said the first law of ecology is "you can't change just one thing," and the same is typically true of institutions.

Institutional Perpetuation as a Personal Matter

Logical consistency alone is not sufficient for the perpetuation of institutions, of course, and we've already seen some of the nuts and bolts that make institutions work in practice. The functionalist paradigm can shed light on how institutions operate.

Through *socialization,* we learn the various components of the institutions we operate within, and we learn that everyone else knows those same beliefs, values, and norms. The sanctions associated with norms are one force for compliance with the rules we've learned.

Socialization often takes place on a subtle level. When my son was born years ago, I was struck by the fact that I basically knew how to be a father—even though I'd never been to father school. Among other things, I knew implicitly that a part of my job was to teach my son what it meant to be a man. More to the point, I recognized that—by example if nothing else—I was teaching him how to be a father, including the part about teaching his son, if any, how to be a father, which would include teaching *his* son . . .

A great deal of what we take for granted in society gets perpetuated through socialization into the agreements comprising our social institutions.

Internalizing those agreements—making them a part of our personal feelings—can be an even more powerful force. If you internalize the view that husbands and wives should limit their sexuality to each other, for example, you are unlikely to violate it. And if you do violate it once, you may feel so guilty that you won't do it again.

Internalization is not limited to "important" matters but includes mundane or commonplace matters as well. My hunch is that you'd feel funny driving through a red light even if there were no cars (or police) for miles around. In fact, if you were the sole survivor in one of those science-fiction stories that kills all but one person on the whole planet, you'd probably still drive on the right side of the road and feel guilty about gassing up without paying.

It is worth noting, rather mundanely perhaps, that as institutions are established, individuals develop *vested interests* in them. Establish a commitment to the importance of seniority within the workplace, for example, and those with seniority are likely to work for the perpetuation of that pattern—and they are likely to have more say in the matter because of their seniority!

Take a minute to consider who would be the most likely to support the perpetuation of these institutionalized patterns:

◆ Tenured faculty have lifetime appointments.
◆ Men earn more than women.
◆ Seniors get better football tickets than freshmen.
◆ Physicians treat other physicians and their families for free.
◆ Women and children leave the sinking ship first.

Just about any institutionalized pattern you can think of is a source of inequality, with the result that those benefiting from it have a vested interest in its perpetuation.

Nothing locks an institution more firmly in place, however, than when individuals link their own *identities* to it. We have already seen that the way we present ourselves to others is a function of the statuses we occupy, and we've also seen how powerfully and implicitly those statuses color our own experience of who we *really are*. The statuses we occupy are inextricably woven into the fabric of institutions.

Suppose you're a mother, for example, and that being a mother is the most important part of your experience of who you are. You look at your children and see proof of your fundamental worth as a human being. Now, imagine someone suggesting that the American family structure should be radically altered—replaced by Israeli-like *kibbutzim,* in which children would be raised in communal nurseries rather than by their parents. I think you can see how threatened you would be by that suggestion. You'd stand to lose something far more important than a lifestyle you enjoyed; your basic sense of *who you are* would be threatened as well. As a matter of fact, just knowing that children are raised in communal nurseries in other societies could represent an assault on your identity. More than one bloody religious crusade has been launched because people somewhere else believed something else.

Thus, when physicians resist changes in the medical care system, more than economic self-interest is involved; their perceptions of *self* may be threatened as well. The same could be said of teachers and education, politicians and government, workers and the economy. To the extent that changing an institution would violate someone's identity, the institution is, to that degree, protected from change. Notice that society is structured in such a way that institutions and personal identities are interwoven just as a matter of natural course.

Institutional Linkages

Sanctions, internalization, vested interests, and personal identity help to ensure institutional perpetuation, and linkages of institutions to one another help as well. I suspect you have some sense of this being true, but you may not appreciate how extensive interinstitutional linkages are.

Let's begin with education. In large part, American schools are funded by government tax dollars. Families send their young to be educated so they can become loyal citizens, productive participants in the economy, and possibly obedient children. Schoolchildren say their pledge of allegiance to the flag and, not all that long ago, said prayers as well. Major religious holidays are recognized by the schools in the form of vacations. Parochial schools, of course, are even more closely linked to religion.

Religious organizations are exempt from paying taxes, and the religious contributions of individuals are tax-deductible as well. Religion, moreover, is a strong proponent of traditional family values, and many American churches are hotly anti-communist and in support of a capitalist economy.

Government and economy are probably more closely linked than any other two institutions; scholars often speak of the "political economy," in fact. Government regulations shape what is possible, probable, and required within the economy, and the economy, by definition, is the source of all wealth— including the revenues needed to pay the costs of government. At a less abstract level, businesses contribute to politicians' campaign funds, and politicians sometimes condition their voting in terms of the interests of a particular company or industry.

I hope these few examples will prompt you to explore the matter further on your own. Do that and you will astound yourself with the extent to which the major institutions of our society are intertwined with one another. You don't need to be cynical in making such an inquiry. Certainly the mutual support of institutions for each other isn't necessarily bad. It would be possible to argue the opposite, in fact. Inter-institutional linkages give the whole society stability. At a personal level, you can experience your life in society as coherent, meaningful, and secure.

At the same time, stability is not necessarily good. Institutional linkages perpetuate the evils of society as well as the benefits. For example, earlier in our history, the slave economy in the South was protected and legitimated by the government—in fact, it was against the law for people to interfere with slavery. Religion provided a moral justification for slavery, and the schools socialized generation after generation to accept slavery.

Thus, stability is neither good nor bad per se. The important lesson of the preceding discussion, however, is that social institutions are designed for perpetuation. Given what we've just seen, you may even be asking how institutions can possibly ever change. That's an excellent question—one we'll now pursue.

Institutional Change

The preceding discussion notwithstanding, institutions *do* change. In fact, change is every bit as fundamental as the powerful forces militating against it. Let's consider some of the reasons why change occurs.

Having just discussed the mutual support that exists among the institutions of a society, we might now observe that such support is never perfect. Institutions also conflict with one another. The beliefs and values of one institution may contradict those of another institution; the values of one may clash with the norms of another. Again, the examples are legion. Here are just a few.

It would be in the interest of the government to collect as much tax revenue as possible, but that would work against the economic interests of businesses and workers. The scientific activities of education often present problems for religion, since they can yield views of reality that contradict traditional religious teachings. Galileo, for example, was forced by the church to repudiate his Copernican view that the earth revolved around the sun instead of vice versa. Similarly, the demands of a career can interfere with family obligations, just as family obligations can prevent or hamper career advancement.

Because institutions are so extensively linked to one another, their conflicts often result in extensive institutional change. For example, governmental actions to prohibit discrimination on the basis of gender (based on the political principle of equality) have resulted in changes in the employment of women in the economy—which, in turn, has produced changes in the family roles of men and women. Many children now grow up in families in which the mother has a career, leading many girls to make different demands on the educational system than they used to. Changes in one institution, then, can ripple across other institutions.

Some institutional change is a product of conflicts within the same institution. Although I've emphasized the degree to which the beliefs, values, and norms of a particular institution are integrated, that integration is never perfect. Racially segregated churches in America always existed in conflict with Christian values of human dignity and love for others. Our constitutional principle of the separation of Church and State conflicts with the granting of tax exemption to religious organizations. All such internal contradictions provide a pressure for institutional change, even though change is by no means inevitable.

Earlier, I spoke of vested interests as a force for perpetuation in institutions, but the same phenomenon can work for change as well. Although the king might grow understandably partial to the perpetuation of a monarchy, the peasants' interests might lie more in the direction of a democratic revolution. And although the capitalists are fond of capitalism, those eking out an existence in the impoverished underbelly of society often find other forms of economy more attractive.

Environmental changes, such as population growth, famine, earthquake, flood, and plague, can also be a source of institutional change. Wars bring about institutional change, as do economic depressions and recessions. Discoveries and inventions are another source of institutional change. Consider the impact of television on the family, education, religion, government, economy, and all other institutions of American society.

Probably no invention of recent decades has had more far-reaching impact than the computer, particularly the development of powerful microcomputers. Consider just a few recent newspaper headlines:

♦ Software Is Transforming Small Business
♦ Computers in Groves of Academe
♦ Computers in Office Change Labor Relations
♦ Computer Dossiers Prompt Concern
♦ Do Word Processors Spoil Writers?
♦ The Spread Sheet Moves into the Executive Suite
♦ Computer as Literary Tool Gains Acceptance
♦ Experts Fear Computers' Use Imperils Government History
♦ The Computer Magazine Glut
♦ Study Finds Computers and Family Mix
♦ Farming with a Computer
♦ Coast Police Use Computer to Clear Area of Prostitutes
♦ Free Speech Issues Surround Computer Bulletin Board Use
♦ Secret Reagan Weapon in Space: Computer
♦ Computer Digests the Talmud to Help Rabbis
♦ Computer Designing Ski Slopes
♦ The Revenge of the Hackers
♦ Concern for Secrecy in Soviet Inhibiting Use of Computers
♦ Computerized Voting Seen as Vulnerable to Tampering
♦ Computer Helps Battle Forest Fires
♦ Computers Aid Study of Ancient Artifacts
♦ For Disabled, Computers Are Creating New Lives
♦ Computer Coding Pursues Fingerprints of Crime
♦ Laser Printer Puts Printing Plant in Home
♦ Computer-Aided Design Dooms Lesser Tools

On balance, then, we've seen a real paradox in the fundamental nature of institutions. On the one hand, they have been designed for survival: sanctions, internalization, vested interests, personal identity, and institutional linkages all conspire to keep everything the same, year after year, generation after generation. If you had been assigned to perpetuate the pattern of social life on a faraway planet, you couldn't have devised a more clever set of mechanisms for that purpose.

Yet, at the same time, we see that social change is inevitable. It is occurring constantly, every day of your life. The rules for living are under continuous amendment.

Here's something even more astounding: *you* are one of the principal agents for both perpetuation and change. Countless times every day you work on behalf of maintaining established social patterns: when you put on clothes, when you stop for pedestrians at intersections, when you refuse to sell secrets to terrorist organizations or have sex with porcupines. At the same time, you are a force for social change every time you stroll naked through the streets or run over pedestrians—not to mention every time you drive faster than the speed limit, talk in the movies, put mayonnaise on your hot dog, or say nice things about your sociology professor. From an institutional point of view, you are both a superpatriot and a traitor.

Whether you abide by institutional agreements or violate them, there's no denying the fundamental power institutions have in our lives. We'll see more of this in the next chapter, which puts all the pieces of society together.

CHAPTER 6

CULTURE AND SOCIETY

In 1968, the ecologist Garrett Hardin published an article in *Science* magazine that is still vital to the study of sociology today.[1] In "The Tragedy of the Commons," Hardin laid out a theme that has persisted throughout this book: the conflict between individual and group interests.

Many years ago, the British "commons"—and, later, the American ones as well—were grassy areas owned by the public. Today's Boston Commons is a park, for example, but in earlier times the British and American commons were used primarily for the grazing of cattle privately owned by individual farmers.

Individual farmers eventually found it was in their *individual* interests to increase the sizes of their herds if possible. Because the grass was free, only a little more work resulted in considerably more milk and meat. As all the farmers increased their herds, though, the commons became overgrazed, and there wasn't enough food for the increased number of cattle. As a result, individual cows produced less milk, weighed less when slaughtered for beef, and eventually a certain percentage starved to death.

It would be worthwhile for you to take a minute to imagine yourself as one of those British farmers. You've increased the size of your herd so as to make a bigger profit, but now the

---

1. Garrett Hardin, "The Tragedy of the Commons," *Science,* 1968, pp. 1243–1248.

overgrazing means you are earning less from each of the cows in your herd. What would you do?

From your individual point of view, it would be intelligent for you to increase the size of your herd further if possible. Even though each cow is less productive than before, having more cows would give you more milk and beef overall. That's essentially what each of the British farmers did, with pre-dictable results. The commons became even more overgrazed, each cow produced less milk or beef, and a higher percentage starved to death. This destructive cycle kept repeating until the commons were finally totally unsuitable for grazing.

Although it's easy to dismiss those old British farmers as stupid, there's a far more important point to be made here. Every time individual farmers increased the size of their herds, it was actually in their *individual* self-interest to do so. Let's see how that works.

Suppose you and I are two farmers grazing our herds on the British commons. Let's say each of us has one hundred cattle and each cow produced an average of $100 worth of milk or beef last year. But because of overgrazing, let's suppose each cow only produces an average of $80 in production this year. That means each of us earned $8,000 this year.

Each of us now evaluates what is happening and makes plans for next year. You, being a responsible citizen, decide it would be inappropriate to further burden the commons, so you keep your herd at one hundred. You figure that $8,000 income next year will be sufficient for you to get by. I, on the other hand, increase my herd by twenty cows to make up for the reduced average productivity.

Because most of the other farmers also increase the size of their herds, the commons become more overgrazed, and each cow only brings an average income of, say, $70 the next year. That means your total income drops to $7,000, whereas mine actually increases to $8,400. You go broke, your family starves to death, and your lineage dies out, whereas my family lives on to destroy the commons altogether.

Although I have perhaps taken some fanciful liberties in portraying this historical episode, the point remains. Very often, that which makes good sense in terms of individual self-interest constitutes a disaster for society as a whole. There is a direct parallel between the tragedy of the commons and the situation of birthrates and overpopulation in the Third World. Ask an impoverished peasant mother why she continues to have child after child, and she will probably answer this way: (1) she needs children to take care of her in her old age, and (2) infant mortality is so high in her society that she must have several children in order to be sure some will survive to adulthood. She's right on both counts. Unfortunately, however, as she and all her neighbors continue to bear baby after baby, their family economies as well as national economies are overtaxed, hunger and poverty worsen, and the infant mortality rate climbs even higher.

In the previous three chapters, we have seen how individual interests can conflict with the demands of groups, organizations, and institutions. Now we are going to take the issue to the limit—to the case of whole societies and their cultures. Let's begin by taking a look at what social scientists mean by the term *culture*.

Culture

In sociological and anthropological usage, **culture** typically refers to the totality of institutional patterns in a society—its beliefs, values, and norms, along with its symbols and physical artifacts. I think it is a useful analogy if you liken the culture of a society to the *personality* of an individual. When we note that Americans differ from the British in many ways, and that both are light-years away from the African bushmen, the differences in question are elements of culture.

A more formal definition of culture—one that is generally respected by social scientists—was devised by the eminent anthropologists A. L. Kroeber and Clyde Kluckhohn.

> **Culture consists of patterns, explicit and implicit, of and for behavior acquired and transmitted by symbols, constituting the distinctive achievements of human groups, including their embodiments in artifacts; the essential core of culture consists of traditional (i.e., historically derived and selected) ideas and especially their attached values; culture systems may, on the one hand, be considered as products of action, on the other as conditioning elements of further action.[2]**

Culture might be regarded as those patterns of living that the members of one society share in common with each other (and with their ancestors in the same society) and that distinguish them from other societies. You might think of culture as providing a "way of life." Importantly, a culture is comprised of institutions, which we examined at length in Chapter 5.

I've already introduced the major components that comprise an institution (and, hence, a culture): beliefs, values, norms, and sanctions. It is worth taking a minute or two to clarify the extent to which our daily lives are constrained by institutional patterns. Consider norms, for example.

Some norms are formally established as laws. We might begin with the garden-variety laws prohibiting murder, robbery, rape, burglary, assault, mayhem, treason, and littering. Speed limits, voting age, drinking age, tax and licensing regulations, and rent controls are also examples of formally established norms. There are laws pertaining to sexual behavior, corporate price-fixing, advertising claims, slander, product safety, and cruelty to animals. If you could know the actual number of local, county, state, and federal laws that prescribe and proscribe your behavior, you would probably be depressed

2. A. L. Kroeber and Clyde Kluckhohn, *Culture: A Critical Review of Concepts and Definitions, Papers of the Peabody Museum of American Archaeology and Ethnology,* Vol. 47, No. 1, 1952, p. 181.

for days. Suffice it to say that you are unlikely to get through the day—any day—without breaking some laws, knowingly or unknowingly.

The laws that govern your behavior, however, are greatly overshadowed by the nonlegal norms of your society. Consider something as seemingly straightforward as eating.

There are norms in our society about what's appropriate to eat (e.g., cow) and what isn't (e.g., cat). Other norms specify what should be cooked (e.g., artichoke) and what should be eaten raw (e.g., lettuce). Some foods should be boiled, some baked, and some fried. There are norms specifying what should be eaten at different times of the day and in what order within a given meal. We agree that certain foods go together (e.g., peanut butter and jelly) and that others don't (e.g., peanut butter and mashed potatoes). Some foods should be eaten with a fork, others with a spoon, and you can pick some up in your fingers—but don't get them mixed up.

I think you can appreciate that I've only scratched the surface of norms pertaining to eating. If you were to dwell on this subject, you might start feeling depressed at all the rules you need to follow in just this one area of life—or you might be impressed at how much you know. Either way, the fact remains that hundreds of norms, probably thousands, dictate how and what you and I are expected to eat in our society.

We could carry out the same exercise with regard to clothing, conversations, shopping, driving, pets, child rearing, dating, hairstyles, work, music, literature, art, worship, hobbies, interior decorating, sports, gambling, Marine Corps boot camp, cooking school, or fistfights—to name only a few areas of life. No matter what you are doing, the people around you have expectations about *how* you will do it.

The sea of norms that you and I share—plus the sanctions, values, and beliefs discussed in Chapter 5—make up our culture. Some parts of our culture are neatly organized and integrated within major institutions (e.g., family, government, education), and others (e.g., standing up for the "Hallelujah

Chorus" of Handel's *Messiah)* don't seem related to much of anything else.

Different societies, of course, have quite different cultures. This can be seen in any and all of the various components we've examined. Let's briefly compare traditional Japanese society and modern American society. In terms of norms, for example, notice how different the earlier discussion of eating would have been if I'd been talking about Japan. Forks and spoons would have been replaced with chopsticks. Raw squid would have figured in somewhere.

Japanese and American values differ. The Japanese, for example, place a higher value on being a team player within an explicit hierarchy of social status. Americans, by contrast, are more committed to individualism, stretching from the frontier rebel mumbling, "A man's gotta do what a man's gotta do," to the free-spirited hippie saying, "Do your own thing," in the 1960s and 1970s.

There are numerous differences of belief. For about 2,500 years, the Japanese maintained a belief in the divinity of the emperor, whereas we Americans tend to believe our president (whoever that happens to be) has all the human flaws we find in ourselves. Although we are relatively polite about Washington and Lincoln, you'll have trouble finding anyone who holds Presidents Clinton, Bush, or Reagan to be divine.

Years ago, sociologists frequently spoke of *national character* and *modal personality* in this regard. Whereas I spoke of the value of individualism in American society, it is an easy jump from that to thinking of the "typical" American as being individualistic; similarly, we may think of the "typical" Japanese as polite, the "typical" Italian as emotional, and the "typical" German as rigid. In recent years, sociologists have devoted less attention to such concepts because they tend to create mistaken views about all the members of a given society being alike.

Although it makes sense to speak of *the* culture of a given society, any complex, modern society actually houses several cultures. Contemporary American society, for example, contains cultures associated with its various ethnic groups: Yankees, Chicanos, Japanese, Jews, Cubans, Chinese, and Portuguese. Sociologists use the term *subcultures* in reference to the cultures of the (sub)groups that make up a society. I hasten to add that subcultures are not homogeneous either; each has its variations.

The danger of stereotyping all the members of a given society or of a subculture is compounded by the common notion that some cultural patterns are better than others. Sociologists use the term **ethnocentrism** in reference to the view that your culture is superior to other cultures. When I mentioned eating raw squid a moment ago, you may have found that a little disgusting. If not, perhaps you could work up some disgust over the notion of eating dog meat or live bugs and worms. There are cultures that include those patterns, and you may find it difficult to see them as simply *different* from ours.

Not all governments are democratic, and it's sometimes difficult for Americans to see monarchies, theocracies, and totalitarian dictatorships as merely different from the way we do things. By the same token, we know that capitalism is only one economic form in the world; yet most Americans would have trouble seeing socialism or communism as equally valid alternatives.

Consider religion. If you have firm religious convictions of your own, you may have difficulty granting equal validity to other religions. If you aren't religious, you aren't immune to this problem: you probably have trouble granting validity to *any* established religious faiths. All these examples offer opportunities for ethnocentrism.

There are two major disadvantages to ethnocentrism: one personal and one social. On the personal level, ethnocentrism is like a set of blinders, shutting you off from new experiences—some of which you might find you enjoy. In this sense, ethnocentrism drastically narrows your experience of living.

On a social level, ethnocentrism lays the basis of intergroup hatred and violence. When people fervently believe that their view of things represents the only Truth, seeing others as misguided, corrupt, or evil, they have fewer compunctions about killing on behalf of their version of righteousness. It has often been observed that religious wars are usually the bloodiest.

Unlike ethnocentrism, sociology promotes **cultural relativity**: the view of cultural differences as merely *different,* not better or worse than each other. The place of cultural relativity within sociology is easily misunderstood, however, and I want to be clear about it with you.

Sociologists do not necessarily *believe* that socialism is as good as capitalism, or vice versa. We do not *believe* that dictatorship is as good as democracy. Individual sociologists may have their own personal preferences, and there are surely patterns of agreement as well. At the same time, however, we recognize that views that hold one form superior to another form blind us from seeing how those various forms function. If you were to begin, for example, with the view that socialism was an evil perpetrated on the world by Satan, you'd have difficulty discovering how socialist economies move goods and services around, interact with educational structures, and so forth. By the same token, you'd have trouble understanding how the nature worship of the Australian aborigines gives identity and solidarity to their tribes if you begin with the view that their religious beliefs are merely stupid superstitions.

If you are willing to approach the study of culture with the view that differences are merely differences, not truths and fallacies, you will be in a better position to discover how societies actually operate. Ironically, you will even be able to determine that some norms, values, and beliefs seem to work better than others. Or you may discover that some differences don't seem to make much difference in that respect. In any event, you can't make these kinds of determinations if you start out with the view that some patterns are intrinsically superior to others.

Cultural relativity is a form of open-mindedness in sociology. However, I mean "open-mindedness" here as an element of our methodology, as one of the tools of the scientific method, not just as a wonderful character trait. I'm not sure sociologists are any more open-minded than other people in that latter regard, but we are duty bound to bring that quality to our scientific investigations.

We often use the term **pluralism** to describe those societies that reflect a broad tolerance for divergent cultures. The United States is sometimes referred to as a "melting pot," representing a blending or amalgamation of cultures into some kind of hybrid. While there is some truth to this depiction, the United States is more accurately described as pluralistic. Look around you and you'll find numerous elements of culture from the various societies of Europe, Latin America, Africa, and Asia, as well as the cultures of the Native Americans—reflecting the varied roots of those who collectively call themselves Americans. You'll also find people still living in subcultures that generally maintain the ways of those root societies.

Society

Throughout this chapter, as well as earlier ones, I've been using the term *society* without defining it—knowing that you have an understanding of the term that accords fairly closely to what sociologists mean by it. Nonetheless, in the remainder of this chapter, we are going to look a little more rigorously at what constitutes a society.

To begin, sociologists often use the term **social structure** in reference to the way statuses are linked to one another in organized human affairs. Clear examples of social structure can be seen in corporate or industrial tables of organizations. Probably the military offers the clearest example of all.

The basic building block of the United States Marine Corps, for example, is the *fire team,* consisting of three riflemen and a fire team leader (see Figure 6.1). The roles associated with each of the statuses comprising the fire team are explicitly detailed in Marine Corps training. The authority of the fire team leader is delineated, as is the way the three rifle-

Figure 6.1
USMC Fire Team

men are expected to relate to one another. The details describing the job of fire team leader include techniques of deployment in combat, how to generate morale, and how to command respect.

Although the Marines say they are looking for a *few* good men, the four-person fire team is only the beginning of USMC organization. Fire teams are organized into squads, with the relationships among that fire teams spelled out in the same kind of detail as those describing the relationships among fire team members. Similarly, the rights and obligations of the squad leader, as well as techniques for fulfilling them, are a part of the social structure of the USMC squad. Squads, moreover, are organized into platoons, platoons into companies, and companies into battalions—as illustrated in Figure 6.2.

Battalions are organized into regiments, regiments into divisions, and three divisions comprise the United States Marine Corps. (I simplified matters by leaving out noninfantry units such as tanks and artillery.) On the whole, the USMC provides a great example of social structure.

Figure 6.2
USMC Organization

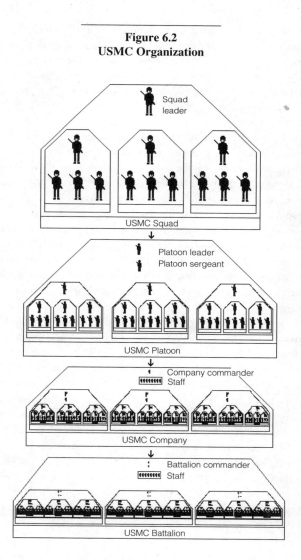

If you find yourself being bothered by this military example, it's a good opportunity for you to practice the kind of sociological detachment that lets you see things the way they are regardless of your personal beliefs, values, and opinions. In fact, if you are deeply committed to world peace, I suggest it would be particularly important for you to know how the military operates.

To recap, social structure includes statuses and the role expectations that link them together. The term also includes norms, sanctions, values, beliefs, organizations, institutions, and the other things we've discussed that "structure" our social life. As you can see, the term *social structure* has a general rather than a precise meaning.

As you'll recall from Chapter 1, sociologists also speak of *social systems*. A social system is essentially a social structure with a *dynamic* quality added. If social structure were likened to an automobile sitting in a parking lot, a social system would be like the same car with its engine running, roaring down the highway.

For our present purposes, I will say a little more about the dynamic quality of social systems. First, systems contain the element of *feedback*. That is, the operation of a system can create changes in the environment within which the system operates and changes in the system itself. Our progressive depletion of nonrenewable resources such as coal and oil is a simple example of a system changing its environment. Population growth is an example of a system changing itself.

The system dynamics method was innovated by Jay Forrester and popularized by Donella Meadows and her colleagues in *The Limits to Growth*[3] and their updated analysis in *Beyond the Limits.*[4] It provides graphic techniques for examining feedback with a system. Consider the simple illustration in

---

3. Donella H. Meadows, et al., *The Limits to Growth,* New York: Universe Books, 1972.

4. Donella H. Meadows, Dennis L. Meadows, and Jørgen Randers, *Beyond the Limits,* Post Mills, VT: Chelsea Green Publishing Co., 1992.

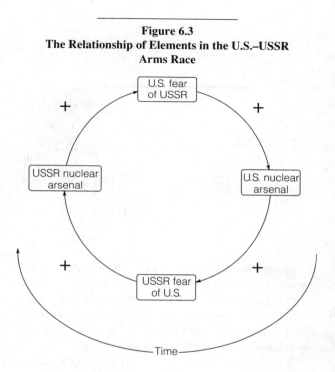

Figure 6.3
The Relationship of Elements in the U.S.–USSR
Arms Race

Figure 6.3 of the nuclear arms race between the United States
and USSR during the cold war.

I've shown four elements in this simple model of the arms
race: the sizes of the U.S. and USSR arsenals and each country's fear of the other's nuclear arms. The model indicates how
those four elements are related to one another in the system.
Let's begin at the upper-left quadrant of the diagram: the
Soviet nuclear arms arsenal and the U.S. fear of that arsenal.
The + sign in the diagram indicates that they were positively
related: if the size of the Soviet arsenal increased, so did U.S.
fears. Notice, on the other hand, that when the Soviet arsenal
was finally reduced, U.S. fears decreased also, another example of a positive relationship.

Figure 6.4
Arsenal Increases in Proportion to U.S.–USSR Fears

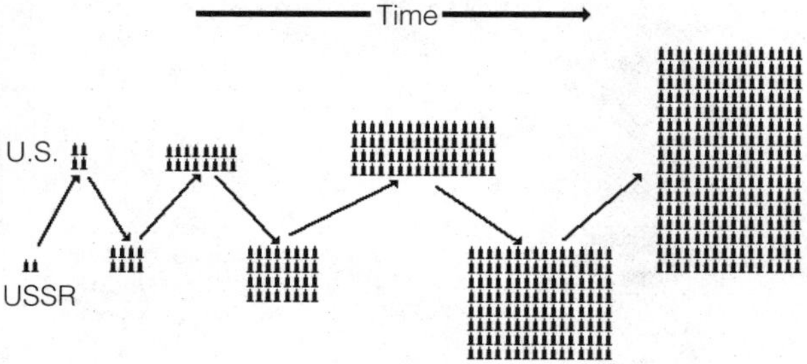

Moving clockwise, what was the relationship between U.S. fears and the size of the U.S. nuclear arsenal? Again, it's positive. As U.S. fears increased, so did the U.S. nuclear arsenal. Now, notice that the bottom half of the model mirrors the top half. When we built more weapons, Soviet worries increased, resulting in more Soviet weapons, leading to . . .

Figure 6.4 shows another way of illustrating the process. In the language of system dynamics, this is a **positive feedback loop**. The term *positive* here does not necessarily mean wonderful. Rather, it means that the system changes in a certain direction, and that change makes the system change even more in the same direction, resulting in further changes, and so on. The wage–price spiral is another example of a positive feedback loop, as is the "tragedy of the commons" discussed at the outset of the chapter. Although positive feedback loops could, logically, be beneficial, they are more commonly dangerous.

While I've presented the relationships among these variables in a schematic format, you should realize that they could also be expressed as propositions in a sociological theory, such as:

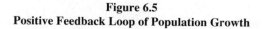

Figure 6.5
Positive Feedback Loop of Population Growth

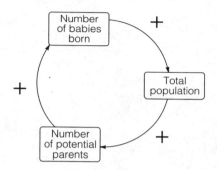

◆ The larger a nation's military arsenal is known to be, the higher the level of fear generated among its adversaries.
◆ The more members of a nation fear the military strength of its adversaries, the more likely that nation will increase its own military arsenals.

Let's look at another system. Consider the model of population growth shown in Figure 6.5. Notice how this qualifies as another positive feedback loop: (1) the more babies born in a society, the larger the total population; (2) a larger population eventually results in more potential parents, which (3) means more babies—and the loop is complete.

This positive feedback loop, however, operates within a larger system that illustrates a **negative feedback loop**, as shown in Figure 6.6. There is a negative relationship between the total population size and the amount of food per capita, assuming a fixed amount of food: the more people, the less food for each. As the amount of food per capita goes down, however, the number of deaths goes up—another negative relationship. Finally, as the number of deaths goes up, the total

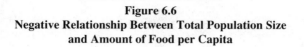

Figure 6.6
Negative Relationship Between Total Population Size
and Amount of Food per Capita

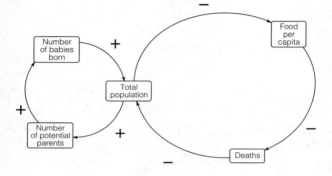

population goes down—also a negative relationship. Notice how this is a negative feedback loop.

1. Let's say the total population *increases*.
2. This means the food per capita *decreases*.
3. Deaths *increase*.
4. This causes the total population to *decrease*.
5. Now, the food per capita *increases*.
6. The number of deaths *decreases*.
7. The total population *increases* (see item 1).

Negative feedback loops create stable systems; sometimes the term *homeostasis* is used in reference to systems that maintain themselves over time. Consider the series of effects in the more complex system:

◆ Total population increases
◆ Food per capita decreases
◆ Death increases
◆ Total population decreases
◆ Number of potential parents decreases

- Number of babies born decreases
- Total population decreases
- Food per capita increases
- Death decreases
- Total population increases
- Number of potential parents increases
- Number of babies born increases
- Total population increases -> Go to top of list.

Even this more complex system is really a simplification of life as we know it. Whereas we have assumed a constant birthrate, it is, in fact, variable. This mediates the relationship between the number of potential parents and the number of babies born. We've also assumed that the amount of food available for the total population was fixed—but it also varies. But how does it vary?

In part, as population increases, the greater demand for food raises its value, thus raising the rewards for growing food and, with it, the actual production of food. Notice, however, that these new factors create a negative feedback loop:

- Decreased food per capita leads to . . .
- Increased rewards for producing food, which leads to . . .
- More food production, which leads to . . .
- Increased food per capita, which leads to . . .
- Decreased rewards for producing food, which leads to . . .
- Less food production, which leads to . . .
- Decreased food per capita -> Go to top of list.

Figure 6.7 shows how the enlarged system works. See if you can follow each step in the system's operation.

These examples should give you some sense of how social systems operate and how you might go about studying them. Before leaving the idea of "system," I want to raise a related issue, even though we won't go into it very deeply.

You may recall from Chapter 5 that the first job of any institution is its own survival. The same could be said about social systems. As with institutions, I don't mean to suggest

Figure 6.7
Relationship among Total Population, Food per Capita,
and Food Production

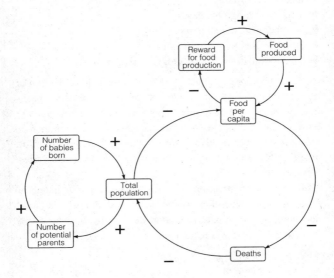

that social systems have humanlike "motivations" or "urges." More simply, I only mean to point out that systems that do not satisfy their own survival needs don't survive—hence, they are not around for sociologists to study. Conversely, those social systems that *are* around for sociologists to study have evidently satisfied their survival needs. Thus, to understand social systems, it is useful to consider what it takes for the system itself to survive.

Without getting into the sometimes complex models various sociologists have worked out, I'd rather just mention a few, fairly obvious examples of system needs. First, for any social system to survive, it must have people. Because all people die, it is useful to see the system need as one of *replacement*. This obviously involves biological reproduction, but the biological function is conditioned by institutionalized norms regarding preferred forms of the family.

Any social system also must have some provision for *leadership*. As we've already discussed, there are numerous options in this regard (e.g., democracy, monarchy, dictatorship), but every social system must handle the need for leadership in some fashion.

There are numerous other system needs that sociologists have pointed to, but I suspect you have enough on this for present purposes. In fact, you should, by now, have a sufficient appreciation for the social systems aspect of our social life.

It has been my observation that as students discover the power of social structure in their lives, there is a tendency to begin seeing society as more orderly and stable than it really is. At this point, therefore, we are going to turn to some of the more unstable—even chaotic—aspects of social life.

CHAPTER 7

INEQUALITY

Several years ago, when I chaired the sociology department at the University of Hawaii, I found myself in a very uncomfortable position. A woman who had tried unsuccessfully for a position on our faculty lodged a complaint with the federal government. Soon, I found myself (representing the department) being charged with discrimination against women.

This predicament totally violated my self-image. After all, I had both spoken and written about equality for men and women. I had served as campaign manager for a woman candidate for city council and was supporting a woman for governor. I certainly didn't feel I had resisted hiring women faculty members, nor did I sense that among the other members of the faculty.

At the same time, there was no denying that only two of the twenty faculty members in the sociology department were women. I was forced to ask myself if that didn't point to a pattern of discrimination over the years. Even if I could feel personally exempt from the charge of sex discrimination during my ten years at the university, could as much be said for the sociology department during its fifty-year history? Prior to answering that, it would be necessary to ask "How would we know?" That wasn't as easy to answer as it might sound.

Ten percent of the department's faculty were women, and that, it was charged, was evidence of discrimination. Perhaps all of us begin with the thought that since half the U.S. population are women, shouldn't half the sociology faculty members be women as well?

The fallacy in this reasoning lies in the fact that substantially fewer than half the sociologists in the United States are women. For example, in 1974, the year in question, only 28 percent of the 632 Ph.D.s earned in sociology went to women. That figure, moreover, represented a substantial growth in the percentage of women in sociology. Five years earlier, only 20 percent of the new Ph.D.s in sociology went to women.

But even if 20 percent was considered an equitable proportion of women, that would mean there should be four, not two, among the department's faculty. However, some of the faculty had been hired much earlier than 1969, during times when the proportion of women in the discipline was even smaller. So what would be the *expected* proportion of women on the faculty if there had been no discrimination over the years? Here's how I decided to estimate that proportion. I think this example will demonstrate some of the complexity involved in matters of inequality in society.

I began with the recognition that each of the faculty members had been selected from among a "pool" of candidates—some men and some women. Each of those pools could theoretically be described in terms of the percentage of women in it. I say "theoretically," since there were no long-term records of who had been considered for faculty positions in the department. It was possible to approximate those pools, however.

For purposes of the analysis, I made the assumption that each of the faculty hired was chosen from among his or her professional *cohort*. That is, when a brand new Ph.D. had been hired—as in my own case—I assumed that the other candidates had also been new Ph.D.s. On the other hand, when a more senior person had been hired, I assumed that he or she had competed with others of the same vintage. Although this is not a completely accurate picture of faculty hiring procedures, it's pretty close to what typically happens.

Since I received my Ph.D. in 1969, I assumed that I had been selected from among a pool of candidates who had the same sex breakdown[1] as all those who received sociology Ph.D.s in 1969 (i.e., 20 percent women). If there had been no discrimination against women, we would expect that the people hired from that pool would be about 20 percent women. Since I was only one person—with one gender—I made a simple bookkeeping notation that the department "should have" hired .8 men and .2 women when I was hired.

I repeated this procedure for each of the twenty faculty members. Another of the faculty, for example, received his Ph.D. in 1972, so I noted that the department should have hired .21 women and .79 men that time.[2]

After making all these calculations, I accumulated the fractions of women that should have been hired over the years, as illustrated in Figure 7.1. The total was three women. That is, if faculty members had been selected *at random*—the ideal model of nondiscrimination, though not necessarily the ideal method of faculty selection—from among the twenty Ph.D. cohorts, it was most likely that the department would have ended up with three women and seventeen men.

Since there were only two women faculty members instead of the expected three, did that indicate a pattern of discrimination? Again, the answer is not obvious, but some standard sociological analytical techniques are appropriate to answer it.

Often in sociological research, we are confronted with empirical observations that may represent a meaningful pattern—or they may simply reflect the vagaries of random chance. That's precisely the kind of question we are now ask-

---

1. In sociology, the term *breakdown* refers to a subclassification of people according to some relevant characteristic(s). Thus the *sex breakdown* means the proportions of men and women. This terminology also allows sociologists to speak of our subjects being "broken down by age and sex"—demonstrating our rapierlike wit.

2. I no longer recall the exact years faculty members received their Ph.D.s, so I have made up the dates in retelling the story. The logic of the analysis is accurate, though, as is the final result of the calculations. Also, the proportions of women Ph.D.s in the years shown are accurate, as reported in the U.S. Department of Education's annual *Earned Degrees Conferred*.

Figure 7.1
Number of Women Who Should Have Been Hired
in the Sociology Department

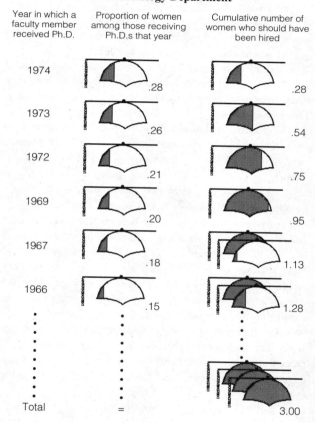

Year in which a faculty member received Ph.D.	Proportion of women among those receiving Ph.D.s that year	Cumulative number of women who should have been hired
1974	.28	.28
1973	.26	.54
1972	.21	.75
1969	.20	.95
1967	.18	1.13
1966	.15	1.28
Total	=	3.00

ing with regard to the possibility of sex discrimination within the sociology department.

Let's take a short detour to illustrate the logic involved here. If you were to flip a coin, you would have a 50/50 chance of whether it came up heads or tails. (Notice that in a single flip, it can only come up one or the other—just as a given faculty member had to be either male or female.)

If you were to flip the coin 100 times, you'd expect to get about 50 heads and 50 tails. I've said "about" 50 of each, since it's unlikely that you'd get *exactly* 50 heads and 50 tails. So, if you were to get 48 heads and 52 tails, that would fall well within our understanding of the probabilities of coin flipping. But suppose you flipped the coin 100 times and got 25 heads and 75 tails. If that happened, you should inspect the coin very carefully, since it is probably weighted in such a way to bias it toward tails.

The question is: how far from the 50/50 expectation does the result have to be to cause us to wonder about the fairness of the coin? Similarly, is the hiring of two rather than three women far enough away from our statistical expectation to cause us to question the fairness of the department's practices?

Suppose faculty members had been selected at random, thereby assuring there was no sex bias involved. There would still be *some* chance that two women would be selected instead of three. Without getting into the details of the statistical calculations, I can tell you that chance is .23; that means there would be nearly one chance in four of picking only two women among the twenty faculty if the selections were made at random. Add the statistical likelihood of selecting only one woman or none at all, and the total probability of selecting fewer than the expected three women faculty members is 40 percent. In other words, being one woman short could easily have resulted from chance and couldn't be taken as conclusive evidence of bias.

A part of the reason for this result is that there are only twenty faculty altogether. By way of comparison, if there were forty members of the faculty and if six women were expected in terms of the logic outlined above, the statistical likelihood of there being four or fewer women drops from 40 percent to 26 percent, even though the expected and observed ratios of women to men stayed the same. In a department of two hundred, with thirty women expected, the likelihood of there being

Table 7.1
The Effect of Faculty Size

Total Faculty	Women Expected	Women Observed	Likelihood
20	3	2 or fewer	.40
40	6	4 or fewer	.26
200	30	20 or fewer	.025

twenty or fewer is only 2.5 percent. The effect of faculty size is summarized in Table 7.1.

Institutionalized Inequality

There's an ironic implication to this effect of size. Suppose for the moment that there were fifty departments in the university, each with twenty faculty; and suppose further that each department should have had three women but only had two.[3] As we've seen, it would be inappropriate to conclude that any one of the departments was discriminating against women. But the same could not be said for the university as a whole. With a total of 1,000 faculty and with 150 women expected, the probability of there being 100 or fewer—purely by chance—is *only 1 in 5,000.* We would not be justified in thinking that the shortage of women faculty members in the university as a whole was merely a result of random chance. But what was it a result of?

Sociologists sometimes use the term **institutional preju-dice** in reference to those situations where prejudice and discrimination appear to exist within a society or large portion of

---

3. Please realize I am using these numbers for illustration only. University departments vary widely in the percentage of women on their faculties, as well as in the percentages that one would expect if there were no gender bias in hiring.

Table 7.2
Median Income of Year-Round, Full-Time U.S.
Workers

Year	Men	Women	Women/Men
1970	$ 9,184	$ 5,440	.59
1975	$13,144	$ 7,719	.59
1980	$19,173	$11,591	.60
1985	$24,999	$16,252	.65
1990	$29,172	$20,586	.71

Sources: U.S. Bureau of the Census, *Statistical Abstract of the United States, 1985,* Washington, D.C.: U.S. Government Printing Office, 1984, p. 452 and U.S. Bureau of the Census, *Statistical Abstract of the United States, 1992,* Washington, D.C.: U.S. Government Printing Office, 1992, p. 452.

it, yet no individuals can be identified as the culprits. Consider the example shown in Table 7.2.

As these data indicate, American women workers earn less than three-fourths as much as men, with some improvement across the time span examined. The pattern of women earning less is a very old one in America. A number of explanations have been given for this.

It is argued that women earn less because they are more likely to work part-time or only part of the year, but these data show the pay difference holds for those full-time, full-year workers as well. It is also argued that women typically occupy lower-status jobs: they are nurses rather than doctors, and secretaries rather than executives. The data show, however, that women earn less when they are employed in the same occupations and professions as men.

In 1970, researchers at the University of Michigan[4] undertook a national survey on occupation and employment to examine in detail the charge that women are discriminated

4. Teresa Levitin, Robert Quinn, and Graham Staines, "Sex Discrimination Against the American Working Woman," Report of the Institute for Social Research, University of Michigan, Ann Arbor, MI, 1970.

against in income. They began their analysis of the data by
selecting half the men surveyed and by studying carefully the
relative importance of different factors in determining their
incomes. They considered occupation, years of experience,
education, and many other relevant factors. Using a technique
called *regression analysis,* they were able to construct a complex equation that they felt would predict the income a given
person might expect to receive based on his or her various
characteristics.

When the researchers applied the equation to the qualifications of the men who had not been used in constructing it, their
income estimates were off by only about thirty dollars a year
(thus reassuring the researchers of the equation's validity).
When the same equation was applied to the qualifications of
the women surveyed, however, the researchers found that
those women actually earned over *$3,000 a year less* than their
qualifications predicted!

It is not without reason that the Census Bureau estimates
that a twenty-five-year-old woman with five or more years of
college will have lifetime earnings of just half that of her
male counterpart; in fact, the twenty-five-year-old woman with
four years of college will earn less than a twenty-five-year-old
male who drops out of high school.[5]

That disparities in the distribution of wealth exist in the
United States occasionally comes as a shock to those who have
not previously examined it. As a background for our continued
examination, Figure 7.2 shows that more than half the nation's
annual income (including capital gains) goes to the richest one-
fifth of the nation's households. The graph also shows the shares
received by the other four quintiles. Notice that the poorest 20
percent of the households receives 1 percent of the income.

Similarly, families in the poorest one-fifth have an average
(mean) net worth of $30,100, contrasted with $586,700 for

---

5. U.S. Bureau of the Census, *Statistical Abstract of the United States, 1985,*
   Washington, D.C.: U.S. Government Printing Office, 1985, p. 453.

Figure 7.2
The Distribution of Income in America, 1990

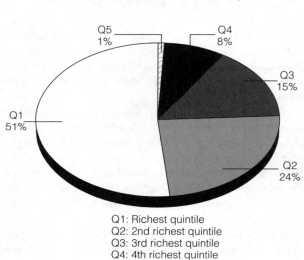

Q1: Richest quintile
Q2: 2nd richest quintile
Q3: 3rd richest quintile
Q4: 4th richest quintile
Q5: Poorest quintile

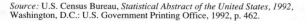

Source: U.S. Census Bureau, *Statistical Abstract of the United States, 1992,*
Washington, D.C.: U.S. Government Printing Office, 1992, p. 462.

those in the top one-fifth. As we saw in our analysis of gender
and income, the differences are not randomly distributed. And
gender is not the only basis of maldistribution.

Consider the income of black and white families in
America, shown in Table 7.3.

As in the case of women, the lower incomes of blacks also
represent a long-term historical pattern—one that goes all the
way back to slavery. Whereas male–female income differences
have been decreasing somewhat, the white–black differences
have not.

The question you might be asking yourself is, "Who's
responsible for these differences?" There's certainly no law
saying that women or blacks must be paid less. In fact, various

Table 7.3
Median Income of Families

Year	Whites	Blacks	Blacks/ Whites
1970	$10,236	$ 6,279	.61
1975	$14,268	$ 8,779	.62
1980	$21,904	$12,674	.58
1985	$29,152	$16,786	.58
1990	$36,915	$21,423	.58

Source: U.S. Bureau of the Census, *Statistical Abstract of the United States, 1992,* Washington, D.C.: U.S. Government Printing Office, 1992, p. 449.

laws prohibit such discrimination. Still, the differences we've just seen are hardly the result of random variations. Patterns exist, and it's worth asking how and why.

Another example of institutional discrimination can be seen in the difference in mortality rates of black infants and white infants (see Table 7.4).

Infant mortality refers to the number of children (per thousand) who die during their first year of life. In 1940, for example, an average of 43.2 out of every 1,000 white children died before they were a year old. For black children that year, the rate was 72.9: more than half again as high a rate.

Although infant mortality rates of both blacks and whites have been dropping steadily—reflecting such factors as improved nutrition, public sanitation, and medical care—the difference between blacks and whites has actually expanded. In recent years, newborn black babies are twice as likely as white babies to die before their first birthday.

The difficulty people have in comprehending the nature of institutional racism was illustrated on April 22, 1987, when the U.S. Supreme Court refused to strike down Georgia's death penalty, despite statistical evidence that it was applied in a racially discriminatory manner. Warren McCleskey, a black

Table 7.4
Infant Mortality Rates in America

Year	Whites	Blacks	Blacks/ Whites
1940	43.2	72.9	1.69
1950	26.8	43.9	1.64
1960	22.9	44.3	1.93
1970	17.8	32.6	1.83
1980	11.0	21.4	1.95
1989	8.2	17.7	2.16

Sources: U.S. Bureau of the Census, *Statistical Abstract of the United States, 1985*, p. 73, *Statistical Abstract of the United States, 1986*, p. 72, and *Statistical Abstract of the United States, 1992*, p. 80, Washington, D.C.: U.S. Government Printing Office.

man, had been convicted by a Georgia court of killing a white policeman and had been sentenced to death. The NAACP Legal Defense Fund appealed his case to the Supreme Court, arguing McCleskey's death sentence reflected a pattern of racial discrimination.

An analysis of more than 2,000 homicide cases in Georgia showed that prosecutors were more likely to seek the death penalty when the victims were white than when they were black, and juries were also more likely to grant the death penalty in the case of white victims. When the victim was white, black murderers were substantially more likely to be given the death penalty than were white murderers: 22 percent versus 8 percent.

Without denying the validity of the statistics, even calling them "troubling," the Supreme Court's five-judge majority said they would require proof of "purposeful discrimination" before eliminating the death penalty on racial grounds. They found no such evidence in the statistical data presented to them, leading one of the defense attorneys to conclude, "The only thing that would fit the bill would be a confession from a juror or a prosecutor, which is obviously not going to happen."[6]

As in the case of women earning consistently less than men, or in the case of higher infant mortality rates among minorities, we are seldom able to point the finger at a clear-cut individual culprit. Yet the sociological point of view requires that we recognize the reality of the pattern and seek its causes. And even when we can identify a Simon Legree or a Hitler, the sociological point of view cautions us to look also for the social conditions that produced that culprit, lest we disempower one evildoer only to find him or her replaced by another.

This is not to say there's no sense in going after identifiable culprits. When there has been a series of burglaries or psychopathic murders, for example, the police have sometimes found the person committing them and have stopped the crimes by locking up the criminal. Unfortunately, when we apply this approach to major social problems, it usually fails.

There are people concerned about economic discrimination against women who will feel they can tell you who is to blame. Some will indict all men, some will lay the problem at the feet of greedy employers, and some will blame politicians. Similarly, there is no lack of suggestions for who is to blame for economic discrimination against blacks and other ethnic minorities.

Sometimes it is possible to identify individual culprits and make them stop. For example, when a chain of health spas in Detroit was found to be paying its male managers a 7.5 percent commission on new memberships, while paying female managers only 5 percent, the Sixth Circuit Court of Appeals made them stop. A Montana bank that was paying female tellers less than male tellers suffered the same fate.

---

6. Reported in David G. Savage, "High Court Rejects Racial Challenge to Death Penalty," *Los Angeles Times*, April 23, 1987, pp. 1, 22. For a more detailed analysis of the court arguments and supporting statistics, see Samuel R. Gross and Robert Mauro, *Death and Discrimination: Racial Disparities in Capital Sentencing*, Boston: Northeastern Press, 1989.

Although actions like these benefited the women involved, women as a category were still earning about sixty-five cents on the dollar afterward. In other words, the larger problem was not solved.

The simple fact is that there is no identifiable culprit responsible for economic discrimination against women, blacks, or anyone else. There is no Archie Bunker masterminding the system from a secret command post in New York City. It's not the work of some secret power bloc or political party. Even if we could actually pin economic discrimination on some individual or group, who's to blame for the black infant mortality rate being twice that of whites?

Some Sociological Insights

When all is said and done, the fact remains that inequality is a fundamental characteristic of social structure. I do not say that to justify inequality or to suggest that those groups who are discriminated against in our society need to be, or even that they are discriminated against in all societies. I merely point out that the structuring of inequality in organized society involves something more complicated than a few bad people doing bad things to others. If we limit our energies to locating the culprits, we will probably accomplish nothing. We may as well try to find out who's responsible for there being religious differences in the world.

This is not to say that we cannot deal with inequality, however. Sociology, although it doesn't have all the answers, offers some useful insights.

To begin, although equality is a basic value in American society, we are not *totally* committed to equality—even in the realm of abstract values. As Seymour Martin Lipset has pointed out, there has been a long-standing conflict between the values of achievement and equality in American history.[7] Although we have a basic commitment to the equal worth of all individuals, we are also committed to being able to achieve inequality. We want to be able to work hard and get ahead. This is one of the fundamental differences we have with the communist model of society.

Inequality, however, is not only a matter of individuals getting ahead of or falling behind others. The most important form of inequality is the organized and persistent **stratification** of large segments of society: men generally doing better than women, for example; whites doing better than blacks.

When you look around the world, you find stratification systems based on various characteristics: race and gender, as we already noted, but also religion, age, family/kinship, occupation, wealth, education, physical prowess, politics, and many others. As I'm sure you realize, different societies value these characteristics differently. As only one example, consider age. In many traditional societies, especially those of Asia, the old have been revered. By contrast, America is commonly referred to as a youth-oriented society, and we often treat our elderly badly.

We often speak of **minority groups** in this context, although a minority group may actually constitute a majority of the population—as in the case of women in America or blacks in South Africa. The term *minority* refers to the group's smaller share of society's rewards: such rewards as money, respect, and political power. Every society seems to have its minority groups: the Tamils of Sri Lanka, the Ainu of Japan, the Kurds of Iran, the Jews of Russia, and the Palestinians of Israel to name a few.

Some sociologists have argued that stratification is inevitable. Four decades ago, Kingsley Davis and Wilbert Moore published a classic statement on the matter.[8] First,

7. Seymour Martin Lipset, *The First New Nation: The United States in Historical and Comparative Perspective,* New York: Basic Books, 1963.

8. Kingsley Davis and Wilbert Moore, "Some Principles of Stratification," *American Sociological Review,* Vol. 10, 1945, pp. 242–249.

Davis and Moore said that some jobs in society seem more important than others; the king or president is always going to rank higher than the street-sweeper. Similarly, some jobs require special skills or special training, suggesting they will reap greater rewards. For these and other reasons, the authors suggested that some kind of stratification is inevitable.

Notice that this view directly contradicts that of an earlier sociologist, Karl Marx, who felt that a *classless society* was both desirable and possible. Although the votes may not all be in yet, there is more evidence to support the inevitability of stratification. If a truly classless society is possible, it hasn't yet been realized. No one would suggest, for example, that a peasant farmer in the former USSR ranked equally with the general secretary of the Communist party. More broadly, the Soviets had to deal with occupational and educational stratification. For example, Dennis Williams reported on the difficulty of getting good students to go into vocational schools:[9]

> **The difficulty in persuading students to switch tracks was evident in a recent article in a Soviet trade paper called Teachers' Gazette. It described a meeting for eighth graders in a Moscow suburb, where a Communist Youth League propaganda specialist urged students to "give their strong young arms and hot hearts where they are most needed." The students barely listened and one, asked if he would go to vocational school, replied: "I don't need any forest ministry," meaning a PTU [vocational school] filled with *oaks,* Russian slang for "dunces." The boy's teacher said in the article that she was unable to recruit 25 of her 42 pupils for vocational school, as the Communist Party wanted: 10 refused outright and 15 more said they would not go to any "dusty" school teaching farm or factory work. The aversion to vocational education is even more pronounced at elite schools like central Moscow's School No. 22, which emphasizes English-language training. There, the students, who are overwhelmingly drawn from the families of diplomats, scientists, and other members of the Soviet intelli-**

---

9. Dennis Williams, "Class Conscious in Moscow," *Newsweek,* June 11, 1984, p. 73.

gentsia, uniformly profess a desire to go on to a university and
follow their parents into a profession or government service.

The Functions of Poverty

In understanding the persistence of inequality, it's important
that we recognize all the ways that some individuals as well as
the society as a system benefit from it. With this in mind,
Herbert Gans looked at the functions of poverty.[10] Here's
some of what he found:

- As long as there are poor people, there will always be
  someone to do the "dirty work" of society.
- Since the poor must accept low wages, the wealthy can
  "afford" luxuries—such as domestic servants—that might
  otherwise cost too much.
- The presence of the poor provides work for those who take
  care of them—ranging from police and social workers to
  "loan sharks."
- Poor people provide a market for inferior goods—such as
  spoiled food—that would otherwise be thrown away.
- The poor can be held up as bad examples to justify and
  strengthen the social norms of hard work, thrift, and the
  like.

Gans is not justifying poverty. He does not suggest that
these "functions" make poverty a reasonable "price to pay."
This kind of analysis need not hold back efforts at social
reform. Indeed, if you were really committed to solving the
problems associated with poverty, then you should recognize
these and other reasons that cause poverty to persist. If we
could somehow end poverty in America, we would need to
make some other arrangements for handling the "dirty work"

---

10. Herbert Gans, "The Uses of Poverty: The Poor Pay All," *Social Policy,*
    July/August, 1971, pp. 20–24.

of society; we'd need to find other work for the social workers; and we'd need to find something else to do with the day-
old bread.

In the preceding comments, I have only touched on a few
elements in the sociological approach to stratification in society. In Chapter 10, we'll return to this topic in looking at stratification among societies: the "haves and have nots" of the
international community. It has not been my purpose here to
give you an overview of even the most important topics in this
field. Rather, I have wanted to demonstrate that we often tend
to approach social problems such as inequality too simplisti-
cally and that we are not very effective when we do so.
Sociology offers insights into the complexities of social life.
Sociological paradigms such as the social systems and the conflict models, for example, offer special insights into matters of
inequality, thus empowering you to exercise your critical
thinking in a way that can make a difference in the world.

CHAPTER 8

FREEDOM VERSUS ORDER

Hans and Emma Kabel don't live in New York anymore. They lived in the same apartment in the Morrisana section of the South Bronx for fifty years. Now they're gone.

When Hans and Emma moved into the Bronx, it was a happy place for them. Weekend trips into the country reminded them of their native Germany. They had good friends and a good life.

Things began changing for the Kabels in the late 1930s when Hans's employer, a meat processing company, moved its operations to Connecticut. Though Hans commuted four hours a day to his job, most of his coworkers moved from New York to be closer to work. That pattern continued until Hans retired from work in 1964.

Financially, the Kabels were reasonably well off. In addition to Social Security, they had Hans's pension and around twenty thousand dollars in savings. In other respects, however, life was less rosy. The Kabels became increasingly isolated, having only a few close friends, and the growing problems of the city began closing in on them.

The South Bronx had become for many an urban jungle. Hostility and violence were everywhere. It had become generally unsafe to be outside, and Hans and Emma increasingly stayed locked in their apartment. When others encouraged them to move—they could afford it—they simply replied that they could not casually walk away from a home they had occupied for half a century.

Then, the violence all around them reached their home. Hans was mugged in the hallway of their apartment building. Even after the muggers took his money, they continued beating him. His head was repeatedly smashed against the marble staircase. Friends said he never recovered mentally.

A year and a half later, violence struck even closer to the Kabels' home of fifty years. Thugs broke into the apartment and tortured the couple in an effort to find their money. When it became apparent that Hans was unable to communicate anything of value to them, the invaders tied and gagged and beat him.

They were more insistent in questioning Emma. Having beaten her severely in the face, one of the thugs began puncturing her arm with a roasting fork, leaving an orderly track of stab marks up her arm until—as he neared her eyes—she gave in and told him where to find two hundred dollars in the apartment. The intruders stayed to terrorize the elderly couple for three-quarters of an hour more.

Those two violent episodes marked the beginning of the end of the Kabels' residency in New York. Clearly they could no longer live in their beloved apartment in safety. Emma began making plans to leave. Then, one day, Emma carefully cleaned and straightened the apartment, wrote letters to relatives in California, and prepared detailed instructions for the Kabels' attorney. Their affairs thus set in order, Hans and Emma Kabel, seventy-eight and seventy-six years old, hanged themselves in the apartment they could not bear to leave in life. Emma's suicide note said simply, "We don't want to live in fear anymore."[1]

Although unusually powerful in its poignancy, the Kabels' tragic experiences are by no means unique. Crime abounds in our society, and there's a reasonably good chance you've been a victim of it. In fact, I'd say it's just about certain that you've

---

1. Details of this story are taken largely from Kenneth Gross's *Newsday* article, printed in the *San Francisco Examiner and Chronicle,* December 12, 1976.

been victimized. If you haven't been personally robbed or beaten, you've probably breathed toxins that were pumped illegally into the air; you've paid prices illegally inflated through price-fixing; you've also paid more for groceries to make up for losses due to shoplifting. Whenever others have cheated on their taxes, you have had to pay more to make up the difference. When others have filed phony insurance claims, your rates have gone up.

In Chapter 6, I pointed out that you and I live in a veritable sea of rules, formal and informal. There are so many, in fact, that it's inevitable we'll violate norms every day of our lives. We'll break some of the rules because we don't know they exist; we break others because we disagree with them; and sometimes we think we're wrong but hope we can get away with it.

See how ironic this situation is: there are rules for just about everything, and people are breaking the rules right and left. Whether you regard society as being overly regimented or too raked with disorder depends on your point of view: similar to whether you view a glass as half full or half empty.

Although some of the rule breaking may seem harmless enough to you, some of it—as in the case of the Kabels—is brutally damaging to others. Ultimately, the issue of law and order needs to be dealt with somehow in any organized society. This involves collective definitions of what is or is not acceptable and mechanisms for enforcing that collective reality.

Unless people basically abide by the norms comprising a social structure, that structure will fall apart, and orderly life together will likely be replaced by chaos. Before you align yourself automatically with the preservation of order, recognize that there are some social structures you might feel should be eradicated. If you were a black person living in South Africa, for example, you might not resonate to the sanctity of law and order. As a general rule, those who are profiting from the system—whatever it is—are inclined to seek its continuation, while those who suffer from the status quo may be, not unreasonably, more interested in seeing it overturned or at least redressed.

This situation is only part of an even more fundamental social dilemma. Social order is a two-edged sword for everyone. Undeniably, all of us—even the most deprived—gain some benefits from society. The most impoverished ghetto resident can be reasonably assured that traffic on city streets will run in the prescribed direction, that the air will be cleaner than unconstrained industrial growth would have produced, and that whatever food he or she can afford to buy from the supermarket will be reasonably pure and untainted. Although we differ in the relative balance of what we put into society and what we get out of it, there is no denying that all of us benefit in numerous ways from the various structures of society.

Yet every one of the benefits we might name comes at a price, measured in individual freedom. The orderly flow of traffic, for example, means we can't drive any way we want without risking punishment. The protection we enjoy against being gunned down in the street comes at the price of not being able to do the gunning down we might feel inclined to carry out from time to time. If examples like this seem a little outlandish, my purpose is to illustrate the inescapable conflict between freedom and order in society.

This is not to say that social structures necessarily deprive us of freedom, however. As I've noted, we often resolve the dilemma by simply breaking the rules. We cannot do so with impunity, however, since any violation of the norms carries the risk of sanctions. And as more people break the rules, those responsible for maintaining order may tighten the screws. Laws may become more strictly enforced, punishments harsher, and so forth. That's not inevitable, but it's commonly the case.

Much of social life, then, can be seen as a contest between deviance and social control. This chapter will examine some aspects of that struggle.

Deviance

What exactly is *deviance?* Most simply put, it is the breaking of rules—any rules. It is the violation of established norms. Murder, for example, is an obvious violation of a rule well established in all societies, although it may be justified under certain conditions. Some acts are deviant in one society but not in another. For example, polygamy is not permitted in the United States, although it is acceptable, even admired, elsewhere. Profit-making is a punishable offense in socialist countries, whereas it's the backbone of free enterprise and entrepreneurship in capitalist countries like our own.

You can be deviant, however, without ever breaking the law. If you're a man, for example, it's perfectly legal for you to wear a dress, but you do so at the risk of substantial ridicule. There's no law that says you have to shake hands when someone offers you that greeting, but you'll be a first-class deviant if you respond by sticking out your tongue—which is, incidentally, the accepted greeting in some other societies.

Sometimes you are deviant by virtue of following the rules. There are city streets marked with a twenty-five-mile-per-hour speed limit where driving that speed will generate honking horns, shaking fists, and other accusations of deviancy. Or spend the day telling everyone the absolute, unvarnished truth, and count your remaining friends at the end of the day.

All of these examples should make one thing unmistakably clear: you are a deviant. You've always been one and you haven't a prayer of escaping your life of deviance. If that's not totally clear, notice that telling the truth can make you a deviant by one standard while lying violates another. There's no way out.

Another essential point needs to be made: just as social structures aren't necessarily good, deviance isn't necessarily bad. Given the fabled big-city norm of "not getting involved," the person who steps forward to stop a mugging is deviant.

Abolitionists broke the law by helping runaway slaves to escape their masters. The Boston Tea Party was blatantly illegal, as was the American Revolution as a whole. Yet we revere all those deviants.

One person's dastardly crime is another person's act of courage and conscience. Indeed, deviance is a prime source of social change, including those changes we later regard as beneficial. All this tends to complicate how you should feel about deviance. It doesn't mean you should honor all deviants—history will probably never look kindly on Hans and Emma Kabel's tormentors—but it does suggest there's wisdom in stepping back and taking a second look before condemning people and their behavior just because they don't fit well with the norms.

Social Control

If deviance is more complicated than it may look on the surface, the same must be said for **social control**—the organized effort to keep deviance in check. Social control runs the gamut from parents, teachers, and peers all the way to courts, police, and prisons. Anything that functions to keep us in line is an example of social control. All institutions, such as family, education, and religion, serve that function.

Whenever the deviance in question is "bad," the social control aimed at preventing it seems "good." But who's to say what's good and what's bad?

During an earlier era in Boston, an unfortunate Captain Kemble was returning home after three years at sea. His wife, understandably excited about her husband's return, came out of the house to greet him. They met, they embraced, they kissed. Warms the cockles of your heart, right? Not so fast. The court sentenced Captain Kemble to spend two hours in the stocks for "lewd and unseemly behavior."[2] The Massachusetts Bay Colony was a bit touchy about s-e-x, but the religious and political leaders were equally sensitive about disrespect for authority. They couldn't keep the lid on completely, however, as Vera Lee points out:

> Still, a surprising number of townspeople did voice contempt for established worthies, and many had to appear in court for picturesque and biting barbs aimed at their political and religious superiors. Ursula Cole, of Charlestown, convicted of reviling the ministers of her parish, declared she'd just as soon "heare a catt meauw as them preach." John Lee, arrested for saying that Governor Winthrop was nothing but a lawyer's clerk, maintained that the court made laws to pick men's purses. And the grand jury indicted George Nubo, of Lancaster, for dubbing the ministers "liars, drunkards, and whoremasters."[3]

Sometimes it seems as though social control has no function except its own perpetuation. Here's an experiment you can conduct on a college campus. Sneak into a classroom used by several instructors and put the following notice on the chalkboard or marker board. See how long it stays there.

Do NOT erase this notice!

Although some of the vagaries of social control can strike us as funny, that's not always the case. More typically, it's anything but funny. Consider a few contemporary examples from the files of Amnesty International:

---

2. Vera Lee, "Crime Among the Puritans," *Harvard Magazine,* July/August, 1986, pp. 48A–48H.
3. Lee, p. 48B.

◆ Kiril Spasov passed his twenty-first birthday last month in a Bulgarian prison. Police arrested him soon after he graduated from high school, and a military court sentenced him in September 1983 to three years' imprisonment for planning to leave the country without permission from authorities.[4]

◆ Eight writers and journalists, arrested by Libyan authorities after they attended a meeting to honor the memory of a poet, are now serving sentences of life imprisonment. A court convicted them in 1980 on charges of contravening a Libyan law that prohibits all gatherings, organizations, and groups "based on a political concept opposed . . . to the principles of the Fatah Revolution." The law in effect bans all political opposition activity.[5]

◆ Raisa Rudenko is reported to have been arrested in Kiev on 15 April 1981 and charged under Article 62 of the Ukrainian SSR Criminal Code with "anti-Soviet agitation and propaganda." This article prescribes up to 12 years' imprisonment and internal exile specifically for the nonviolent exercise of the right to freedom of expression if the views expressed are disapproved of by the authorities. [Specifically, Mrs. Rudenko agitated for her husband's release from prison and illegally received letters from him.][6]

◆ AI has learned that Tshisekedi wa Mulumba and Kanana Tshiongo were tried by Zaire's state security court on January 10, 1986, on charges of insulting the Head of State. On January 17, 1986, the court announced its verdict: both men were convicted and sentenced to 18 months' imprisonment and to large fines. . . . Kanana Tshiongo was convicted of making derogatory remarks about President Mobutu at the time of Tshisekedi wa Mulumba's arrest. The two men have no right of appeal.[7]

---

4. *Amnesty Action,* published by Amnesty International, April, 1985.

5. *Amnesty Action,* April, 1985.

6. Amnesty International case file.

7. *Urgent Action* notice, Amnesty International, March 1, 1986.

Let me quickly summarize where we've come so far. Social norms are organized so as to make it possible for individuals to live together in relative peace and security. Each of us gains that peace and security by giving up freedoms we might otherwise have. This system cannot work, however, unless people abide by the established norms, and social control mechanisms are created to ensure obedience to the norms. Two potential problems arise: (1) the norms may be unjust or inappropriate (by some standard) and (2) the agents of social control may become more committed to the survival of the system (and their authority within it) than to the purposes for which the system was created in the first place. Now let's look at a third problem.

Even when we might agree that the norms are just and the authorities honorable, our systems of social control simply don't work. Consider just two examples from the area of punishment.

When prisons were first established as a form of punishment, they were generally regarded as a progressive development. Previously, criminals were only incarcerated as an insurance that they'd stick around for their trials and subsequent punishment— often mutilation, torture, or death. Just locking people up with free room and board as their punishment was undoubtedly seen by many as "bleeding-heart mollycoddling."

We never have been completely clear or agreed about the purpose of prisons, and we are even less clear or agreed as to whether that purpose was being served. Four purposes are commonly mentioned.

◆ *Punishment:* Because criminals have caused suffering to others, it can seem morally just that they be made to suffer in return.

◆ *Protection of society:* Locking criminals up keeps them from committing crimes against the rest of us.

◆ *Deterrence:* The threat of imprisonment will keep people from committing crimes.

◆ *Rehabilitation:* Criminals should be resocialized so they will lead law-abiding lives upon their return to free society.

How do you suppose we've done in each of these respects? On the whole, prisons seem to succeed as punishment. Even though some people may live more poorly on the outside than some prisoners, prison life is generally not to be envied.

Similarly, prisons are somewhat successful in protecting society, by keeping criminals away from the rest of us. This only works while they are in prison, however. And some prisoners return to free society resentful of their punishment and better trained in crime, having learned from other inmates.

There is little to suggest that prisons have much of any effect as a deterrent to crime. In fact, those who are most familiar with how bad prison is—former convicts—are the most likely to commit crimes that warrant their being sent back for more. I'll have more to say about deterrence in a moment when we look at capital punishment.

Finally, a historically more recent aim of prisons has been that of rehabilitating prisoners—resocializing them so they can be more law-abiding, productive members of society. I think all of us today would like to see prisons serve this function, and prison administrators have tried numerous programs to realize that purpose. In 1975, Douglas Lipton, Robert Martinson, and Judith Wilks published a review of 231 research reports evaluating a variety of rehabilitation programs, an analysis requested by the New York Governor's Special Committee on Criminal Offenders.[8] Their overall finding: nothing works! None of the evaluations pointed to a program that substantially reduced recidivism rates.[9]

As you can imagine, this conclusion generated a good deal of controversy—and the debate continues. On the whole, however, you would be hard pressed to demonstrate that prisons are effective at rehabilitating people. It has been said that the only thing prison cures is heterosexuality.

---

8. Douglas Lipton, Robert Martinson, and Judith Wilks, *The Effectiveness of Correctional Treatments,* New York: Praeger, 1975.

9. *Recidivism* is a general term referring to ex-convicts getting in trouble again: measured variously as being arrested, being convicted, returning to prison, etc. As a rule of thumb, about 60 percent return to prison.

Table 8.1

Average Murder Conviction Rates (per 100,000 Population)
in States Using Capital Punishment and Those Not Using It

	Capital Punishment Not Used		Capital Punishment Used	
	1967	1968	1967	1968
First-degree murder	.18	.21	.47	.58
Second-degree murder	.30	.43	.92	1.03
Total murders	.48	.64	1.38	1.59

Source: Adapted from William C. Bailey, "Murder and Capital Punishment," in William J. Chambliss (ed.), *Criminal Law in Action,* New York: John Wiley, 1975.

Let's consider the issue of capital punishment—the death penalty. This is another highly controversial issue. We might begin by noting that capital punishment is justified on essentially the same grounds as imprisonment, with the exception of rehabilitation. It's justified as fitting punishment: an eye for an eye. Also, although people disagree on this moral side of the issue, there's no denying that it protects society from any further harm by the criminal. But what about the purpose of deterrence?

Sociology has nothing to say about the issue of morality, but it does offer ways of analyzing the deterrent effect.

Note that some states allow the death penalty for certain crimes (such as murder) while others do not. If the threat of capital punishment were a deterrent, we might expect that states with the death penalty would have lower murder rates than those states that don't have it. Let's look at the facts.

Back in the 1960s, when capital punishment was being widely used in America,[10] William C. Bailey analyzed the murder rates of states that had the death penalty and those that did not. Bailey's findings appear in Table 8.1.

10. The U.S. Supreme Court ruled against the death penalty in 1972, having concluded that it was applied disproportionately against blacks. Since then, however, a few states have rewritten their capital punishment laws in such a fashion as to be acceptable to the Court.

At first blush, these data certainly do not suggest that the death penalty works as a deterrent to murder. In fact, the states with capital punishment had higher murder rates than those without it. Bailey went beyond these initial findings, however, to test another possibility.

Perhaps the direction of the relationship between the death penalty and murder rates was just the opposite of what we've been assuming. Perhaps states with high murder rates tended to institute the death penalty as a reaction, and those with low murder rates tended to repeal the death penalty as unnecessary. If that were the case, then it would still be possible for the death penalty to have its deterrent effect, even though the over-all comparison of rates didn't support the idea.

To test this possibility, Bailey examined changes in the murder rates that might be related to changes in capital punish-ment within given states. If a state abolished its death penalty, for example, we should expect to find an increase in its murder rate subsequently; and states instituting the death penalty should enjoy declining murder rates. Bailey's analyses, how-ever, showed no changes in murder rates that could be associ-ated with changes in whether a state had capital punishment.

The deterrent effect of various punishments is still not a settled issue within sociology. It's a much more complex mat-ter than I've indicated here, but my purpose for now has been to point toward the complexity that exists rather than detail it all. I trust I've shown that sociology offers ways of looking more deeply into issues than is common in everyday "common sense" discussions.

Deviance, Freedom, and Responsibility

I will close this chapter by raising another issue in regard to the tension between freedom and order. Possibly one of the most powerful impacts sociology has had on our social life lies in the context of crime and punishment. In the process we have perhaps solved one problem and created another. I'll close the discussion by pointing to a larger context within which the issue exists.

Let's begin with the question, "Why do people commit crimes?" Years ago, there was no question in anyone's mind about the causes of crime: bad people do bad things. People used to talk about "bad blood," and criminality was seen as partially genetic. From some religious perspectives, criminality was seen as the work of the devil.

Sociology has offered a very different explanation: criminality and other forms of deviant behavior can often be explained in terms of social environmental factors such as poverty, prejudice, lack of employment opportunities, subcultural norms, and disasters. In short, it is possible for a perfectly good person to go bad through no fault of his or her own. People sometimes say that a person was driven to crime by forces beyond his or her control.

I think the political and general public acceptance of this sociological view has been an improvement on the earlier views of deviance. At the very least, it opens up possibilities for rehabilitation and takes account of mitigating circumstances where appropriate. More importantly, it opens up new avenues for the prevention of crime—addressing the root causes rather than appearances.

These advantages, however, have come at a price: the discounting of individual responsibility. Although it is humane and compassionate to understand that the unemployed and impoverished ghetto dweller, for example, was driven into crime by his or her environment, this view implicitly says that person is inferior to those environmental forces—rather like a puppet. Ironically, this compassionate view of human beings is also a demeaning one.

This dilemma, moreover, is not limited to matters of deviance: it is a fundamental one within sociology. The most general contribution of sociology to our understanding of human life is to show how social structures and processes influence individual behaviors, characteristics, and situations. Many things about you, for example, are directly drawn from the family you grew up in: your social class, for example, or perhaps your religious or political views.

In demonstrating the impact of such deterministic mechanisms, however, this sociological view seems to downplay the options available to individuals as well as their accountability for the options they choose. Yet, notice how implicit this view is in the attempt to understand human behavior.

When we ask, for example, "What are the causes of prejudice?" we assume implicitly that prejudice has causes—such as lack of education, prejudiced parents, political or religious views, childhood experiences, and economic competition. We never say that someone is prejudiced simply because he or she wanted to be. Instead, we look for social causes—and we find them. We are, in fact, able to explain things like prejudice on the basis of factors that the people involved have no control over. But notice that every time we find such an answer, we implicitly deny a degree of individual freedom (and responsibility) to those we have explained. By extension, logically, we reduce our own freedom.

Thus, as we've seen, the notion of individual freedom conflicts with order in two ways. On the one hand, social order can only be created through the reduction of freedom: we all have to give up some degree of freedom in order to live together in reasonable harmony. At the same time, and more profoundly, the search for order in our understanding of human behavior mounts an even more subtle attack on the idea of freedom.

CHAPTER 9

SOCIAL CHANGE

The term *social change* typically brings forth images of profound, long-term changes in the structure and process of human social life: the rise of agriculture 9,000 years ago, the Industrial Revolution, the explosion of mass democracies in the latter eighteenth century. As powerful as these examples are, it is by no means necessary to look that far back for important social change.

In particular, the geometric evolution of technology offers numerous examples closer to home. Some people remember the arrival of the automobile as a practical, family vehicle. Many can recall life before television. These two inventions have had radical impacts on almost all aspects of social life.

Even if you aren't old enough to remember the advent of automobiles or television, it is worth taking a minute to recall the technological developments that *have* occurred during your lifetime—and imagine how different your life would be without those innovations. If you are a woman, consider the impact on your life of a growing sexual equality. Or consider the impact on your life if you are *not* a woman. It affects your employment opportunities, and it affects your day-to-day interaction patterns. How has your life been affected by the end of the cold war? Think about changes in shopping, employment, travel, and other aspects of your life.

Consider the arguably most important technological development of our time: the computer. If you are old enough to be reading this book, the computer has changed dramatically dur-

ing your lifetime, and those changes have had a potent influence on your life. I know they have influenced mine.

My first "hands-on" contact with computers came when I
was a graduate student at the University of California at
Berkeley, working as a research assistant at the Survey
Research Center. In those days, the Center was located in a
former fraternity house on the west side of campus. Its ivy-
covered brick exterior created an impression of quality, even
elegance, that overstated matters for the hand-me-down office
furniture and worn carpets inside.

There was one striking exception to this generally funky
motif. The "Machine Room," just inside the front door, occupied the largest room in the building. Back in the days when
air-conditioning was still fairly rare, the Machine Room was
the only part of the Center so blessed.

While there were a number of machines in the Machine
Room, the undisputed centerpiece was the IBM 1620 computer. If nothing else, it was bigger than any of its companions.
Though it could read punched cards, much of its data handling
was done on two magnetic tape drives, each the size of a full-
size refrigerator or soft-drink machine. Its high-speed printer
was the size of an old-time rolltop desk, and its central processing unit and control panel could have inspired science-fiction movies about nuclear power plants. It actually had colored
lights that blinked as the computer "thought."

A rotating staff of "machine operators" ministered to the
1620's needs on a twenty-four-hour schedule. Theirs was a
truly arcane knowledge, and I wouldn't have tried to take
charge of that computer any more than I'd shove a priest aside
and try my hand at saying mass. Although I eventually was
able to mount my own magnetic tapes on the tape drives, I
didn't dare do that without a machine operator standing by.

The point of this lengthy reminiscence has been to illustrate
the incredible evolution of computers in just a few years. I've
been writing this chapter on a small, battery-operated, "lap-
top" computer while sitting in a hotel lobby. My Macintosh
Powerbook is smaller than many cities' telephone directories.

The little computer I carry aboard a plane in my briefcase, powered by rechargeable batteries, has 167 times the computing capacity of the IBM 1620. Whereas the IBM 1620 served the research needs of thirty to forty active social researchers with 24,576 *bytes* of random access memory[1] (24K RAM), the lap-top at my fingertips now has more than four megabytes (4 million bytes) of RAM, and it could be expanded to hold more.

Whenever circumstances make it inconvenient to carry even the lap-top with me, I depend on a Poqet PC, about the size of three decks of playing cards and powered by two AA batteries. With 640K of RAM, it is still more than twenty-six times as powerful as the IBM 1620.

Whenever I have written something that I'm especially proud of or something I want comments on, I can send it to my publisher or to colleagues around the world, using telephone lines to transmit book manuscript. Once a book manuscript is complete, I can dump it onto a 3.5-inch disk and send it to my publisher.[2] They'll make editing corrections on the disk, and that same disk will be used by the typesetter to prepare plates for printing.

To complete this techno-tale, by the time you have this book in your hands, much of the technology I've discussed will be outdated. Save the book for your grandchildren, and they'll laugh hysterically about the primitive "olden times" when you grew up.

This discussion of technological change has a sociological counterpart in social change. Just consider the social changes brought about in your life by computers, whether you own one or not. Or consider social changes that seem pretty far removed from technology. Tastes in music seem to change at least once per generation, for example. Monarchies and aristocracies are replaced by democracies. Religions rise and fall,

---

1. A byte is the memory space required to store a single character, such as the letter *a*.

2. It was a sobering experience when I realized that I now spend most of my working day rearranging tiny magnetic charges on a plastic disk. On my best day ever, all those little charges put together wouldn't fill a gerbil's eyedropper.

and some evolve over time. A number of societies have seen the extended family replaced by the nuclear family. Divorce is no longer the hushed-tones scandal it was in my youth, and recently in America, intentional single-parent families have become more acceptable. As a whole, Americans have become more tolerant of gays and lesbians, as well as religious and ethnic minorities. Abortion, while still controversial, is acceptable to the majority of Americans. The list is endless.

Whenever you get involved in the intricate details of social structure—norms, sanctions, statuses, and all that—it's easy to drift into the belief in the immutability of established institutions. In fact, society is every bit as fundamentally a matter of change as of stability. That's what I want to look at in this chapter.

Kinds of Social Change

Some social change, such as fads, fashions, and crazes, is transitory. Hemlines and hair lengths go up and down like yo-yos, which rise and fall in popularity themselves. Dances come and go: the minuet, the Charleston, the jitterbug, the twist, the frug, disco, the slam dance, and the Achy Breaky, to name just a few. Like a wave sweeping across an expanse of ocean, entertainers rise to prominence and then recede into the indistinguishable mass. Our use of leisure time seems especially susceptible to passing fancies. Some changes come to stay, however.

A recognizable form of baseball was being played on the village greens of New England as early as the 1820s.[3] It was

---

3. My information on the history of baseball is taken solely from Stuart Berg Flexner, *Listening to America,* New York: Simon & Schuster, 1982, pp. 32–47. I tell you this because I know baseball fans are sometimes touchy about the fine points of baseball history, and I don't want to take the heat for someone else's views. If you have any complaints about anything I've written on this topic, Simon & Schuster is located at Rockefeller Center, 1230 Avenue of the Americas, New York, NY 10020. (Call before you show up.)

sometimes called *town ball* in recognition of the fact that town residents would form a team to compete with teams from other towns. The game gained a good deal of national popularity during the Civil War.

Then, in 1867, the Rockford, Illinois, team instituted a change that was to have a lasting impact. To beef up the local volunteers, the team *paid* a few expert players to join them. Although only a few players were paid, and they were paid for only part-time work, the die was cast. In 1869, the Cincinnati Red Stockings was a fully professional team; in a national tour, playing town teams from the Atlantic to the Pacific, they didn't lose a game. Within two years, professional baseball teams were being organized all across the country. Although unpaid town teams can still be found in rural pockets of America, baseball today is more a matter of million-dollar contracts, television, trading cards, and underarm deodorant endorsements.

Without wishing to offend any baseball zealots (nobody needs that kind of enemy), I should point out that sociologists typically are interested in more substantial social changes. Consider, for example, the sweeping, global shift from rural to urban societies. The German sociologist Ferdinand Toennies coined the terms *gemeinschaft* and *gesellschaft* to contrast the face-to-face intimacy of a small town with the largely impersonal quality of a large city, respectively.[4] Thousands of other sociologists have focused on various aspects of this broad social change.

Demographers study trends in population characteristics. Some populations grow rapidly, eventually slowing their growth to the point of population stabilization (zero population growth). Important advances against childhood diseases may result in a lowering of the average age in a society, significantly enlarging the young, dependent population, which must be supported economically by those who work. By contrast, as

---

4. Ferdinand Toennies, *Community and Society*, trans., Charles Loomis. East Lansing: Michigan State University Press, 1957 [1887].

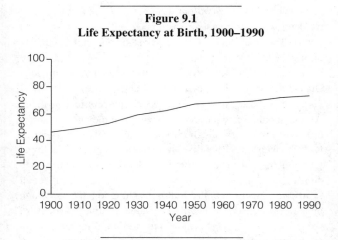

Figure 9.1
Life Expectancy at Birth, 1900–1990

Sources: U.S. Bureau of the Census, *Statistical Abstract of the United States, 1985,* Washington, D.C.: U.S. Government Printing Office, 1985, p. 69; U.S. Bureau of the Census, *Statistical Abstract of the United States, 1992,* Washington, D.C.: U.S. Government Printing Office, 1992, p. 76; U.S. Bureau of the Census, *Historical Statistics of the United States,* Washington, D.C.: U.S. Government Printing Office, 1960, p. 25.

populations cease growing and stabilize in size, the average age tends to get older—which enlarges the elderly, dependent population, which must be supported economically by those who work. As you can imagine, these kinds of changes have impacts on many sectors of society.

Some social change is relatively *linear:* it progresses steadily in a given direction. Consider, for example, the increasing life expectancy of Americans during this century: going from 47.3 years of age in 1900 to 75.4 years of age in 1990 (see Figure 9.1).

Some social changes increase more than linearly. Look at the history of population growth during our national history (shown in Figure 9.2).

This kind of "J curve" is not uncommon in human affairs. Consider a variable closely related to population growth: the number of registered motor vehicles. According to the Census

Figure 9.2
United States Population Growth, 1790–1990

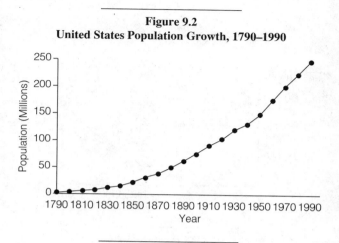

Source: U.S. Bureau of the Census, *Statistical Abstract of the United States, 1992,* Washington, D.C.: U.S. Government Printing Office, 1992, p. 8.

Bureau, there were eight thousand cars, trucks, and other regis-tered motor vehicles in 1900, when such records began. Ninety years later, there were over 189 million. You'll notice in Figure 9.3 that there is a leveling off of motor vehicles during World War II, but other than that, the accelerating growth is steady.

As the number of motor vehicles on the highways has increased, so have the number of highways. Figure 9.4 shows the miles of paved highway in the nation during this century. If you look carefully, you'll see that this graph somewhat resem-bles a backward, flattened letter *S,* and it is called an "S curve." The growth in highways increased gradually until about 1930; the next thirty years saw an accelerated growth, followed by a little tailing off since 1960.

The significance of the slight difference in the patterns of growth in the number of motor vehicles as contrasted with the miles of paved highway can be seen in Figure 9.5, which offers an indicator of highway congestion. Here, I have pre-

Figure 9.3
Registered Motor Vehicles in America

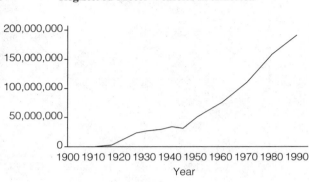

Sources: U.S. Bureau of the Census, *Statistical Abstract of the United States,*
1985, Washington, D.C.: U.S. Government Printing Office, 1984, p. 597; U.S.
Bureau of the Census, *Historical Statistics of the United States,* Washington,
D.C.: U.S. Government Printing Office, 1960, p. 462; U.S. Bureau of the
Census, *Statistical Abstract of the United States, 1992,* Washington, D.C.: U.S.
Government Printing Office, 1992, p. 606.

sented the number of registered motor vehicles per mile of
paved highway. Through 1905, the miles of highway outnum-
bered the number of vehicles. By 1910, the balance had tipped
with two registered motor vehicles for every mile of paved
highway in the country. This increased dramatically until
World War II and resumed even more rapidly afterward. In
1990, the Census Bureau reported there were fifty-four registered motor vehicles per mile of paved highway.

It's important to recognize the *microsociological* implications of these *macrosociological* analyses. These are not merely statistics. They are the cars sitting ahead of you in a traffic
jam. They are the time it takes you to get where you are going.
They are the accidents you drive by or participate in.

I hope these few graphs will give you some ideas about
other examples of social change that you could examine. You

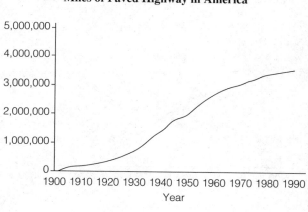

Figure 9.4
Miles of Paved Highway in America

Sources: U.S. Bureau of the Census, *Statistical Abstract of the United States, 1985,* Washington, D.C.: U.S. Government Printing Office, 1984, p. 591; U.S. Bureau of the Census, *Historical Statistics of the United States,* Washington, D.C.: U.S. Government Printing Office, 1960, p. 458; U.S. Bureau of the Census, *Statistical Abstract of the United States, 1992,* Washington, D.C.: U.S. Government Printing Office, 1992, p. 602.

might want to try your hand at it. All you need is a source of data (such as the *Statistical Abstract)* and some graph paper.

We've also begun to see some of the possible variations in the patterns of social change. I suspect all of us have an implicit notion of social change occurring in a fairly steady, linear fashion. As these few graphs have indicated, there are other possibilities. We've only scratched the surface.

Figures 9.6 and 9.7 show two other patterns of social change that take a somewhat longer-term view. These are idealized models; reality would never be as smooth as these models suggest.

Figure 9.6 reflects the fact that major stages in the evolution of civilization proceed in *steps* rather than in a steady increase. Thus, the creation of vast agrarian societies around 9000 B.C. represented a radical shift in the nature of social life: societies

Figure 9.5
Number of Motor Vehicles per Mile of Paved Highway

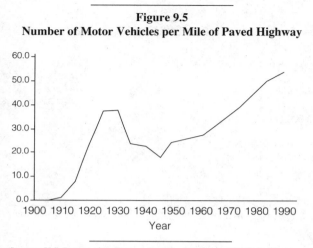

Sources: U.S. Bureau of the Census, *Statistical Abstract of the United States,*
1985, Washington, D.C.: U.S. Government Printing Office, 1984, pp. 591, 597;
U.S. Bureau of the Census, *Historical Statistics of the United States,*
Washington, D.C.: U.S. Government Printing Office, 1960, pp. 458, 462; U.S.
Bureau of the Census, *Statistical Abstract of the United States, 1992,*
Washington, D.C.: U.S. Government Printing Office, 1992, pp. 602, 606.

got a little bigger and less nomadic. Similarly, industrialization
was not merely the continuation of trends in feudalistic societies—it marked the death of feudalism. Notice, this model is
reminiscent of an earlier discussion of paradigm shifts.

Some sociologists have been particularly interested in what
they perceive as *cyclical* social change. For example, Pitirim
Sorokin, an immigrant to America following the Russian
Revolution, felt that three points of view characterized societies from time to time (see Figure 9.7). The *sensate* point of
view defines reality in terms of sense experience. Science is an
example of that point of view. The *ideational* point of view, by
contrast, places greater emphasis on spiritual and religious factors. Whereas Auguste Comte, the father of sociology, saw a
linear evolution from religion to science, Sorokin suggested
that these two points of view alternated cyclically in societies
over time. Finally, Sorokin's third point of view, the *idealistic*

Figure 9.6
Types of Civilizations

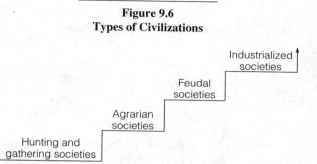

point of view, combined elements of the sensate and ideational in an integrated, rational view of the world.[5]

These are just a few of the social change models that sociologists and other scholars have devised as ways of representing the nonstable aspects of social life. I trust that these examples may suggest other possibilities to you.

Conflict and Change

It is inevitable, perhaps, that our discussion of social change has dealt so much with technology. Indeed, technological developments are a major source of social change. That's not the whole picture, however, and I will conclude this chapter with a brief examination of the function of conflict in social life. There are aspects of social conflict, I suggest, that are not at all obvious to you.

I suspect that most of us regard conflict as fundamentally bad, representing, as it does, everything from arguments and

---

5. You can learn more about Sorokin's model in his monumental four-volume *Social and Cultural Dynamics,* New York: American Book Co., 1937–1941.

Figure 9.7
Sorokin's Model of Cyclical Change

fistfights to revolutions and riots all the way to world wars and nuclear holocausts. It's difficult to find poems written to honor such things. What humane being would choose strife over harmony, war over peace, hate over love?

At the same time, we have already seen situations in which conflict is, even in the eyes of reasonable and humane people, more appropriate than harmony. Many would argue there's nothing admirable about "going along" with injustice, for example. Slave revolts, the civil disobedience of Gandhi, and our own American Revolution all qualify as examples. In fact, if you think about it, there are a lot of poems and songs about revolutions and other forms of social conflict.

We might say, therefore, that there will always be social conflict—and the attendant social change—as long as there is injustice, but whose injustice? Isn't injustice, like beauty, in the eye of the beholder? More accurately, then, we might say that there will always be social conflict as long as there is perceived injustice.

None of this is to say that all social conflict is noble and wonderful. Some flows from greed, bigotry, megalomania, and other negative human traits. Thus, some conflict arises whenever one person (gang, nation) tries to take something from another. Without a question, therefore, there will always be conflict and the potential for profound social change as a consequence.

It is worth recognizing that the social change that results from even "negative" conflict is not necessarily bad. For example, Hitler's bigotry and nationalistic ambitions prompted

World War II and the deaths of tens of millions around the globe. Among the many other social-change consequences of the war, Americans in general became less parochial, coming to recognize the diversity of cultures in the world. Black Americans, sent to die for their country, came home with an irreversible unwillingness to be second-class citizens. The war similarly disrupted the conventional role of women in America, sowing the seeds of equality. There are any number of other "positive" social changes that can be traced to our national experience of World War II.

It is important to realize that this kind of sociological analysis neither constitutes nor implies a justification for wars. It is not to say that, on balance, World War II was worth it. Rather, it is merely a recognition that social life is a complex fabric—every negative has positive aspects, just as every positive has negative aspects. To fail to see and acknowledge this complexity is to remain partially blind to the realities of society and hence to become victims of its functioning. We can never take effective charge of society until we are willing and able to understand it. Specifically, we cannot simply turn aside from conflict as if it were always and totally negative.

As I write about social conflict, I know you will recall our earlier discussions of this. Conflict theory is one of the major theoretical points of view used by sociologists, contrasted to interactionism and functionalism. It is possible to see all social life as a struggle for domination, just as we can see it all as a process of interaction among individuals or as a structuring of social statuses.

A common difficulty for beginning students in sociology is represented by the question: "Which of those points of view is the *right* one?" The proper response to that question is: "None is right or wrong; they are just different points of view that may be more or less useful in making sense of our various observations of social life." I am going to conclude the present discussion by seeming to contradict that sociological wisdom.

There is an important sense in which conflict is the most fundamental aspect of our social life, more fundamental than interaction or structure. It is easiest to see this from the point of view of the individual, but I suggest that what I am about to say has its counterpart at the group level.

As an individual, social conflict can be seen most fundamentally as my *resistance* to whatever is around and outside me. If you want me to be your slave and I resist, that's an easy example of conflict between us. If you express an opinion and I disagree, that's a subtler form of resistance on my part and hence conflict. If I disagree and don't say anything to indicate my disagreement, that's subtler still, but it is an instance of resistance and conflict nonetheless.

Now, if you can further expand your sense of what I mean by resistance and conflict, you'll be able to see that simply *being different* is a subtle form of the same thing. One of the formal definitions of conflict, in fact, is "to be at variance," as in the "conflicting accounts" offered by witnesses to an event. Thus I conflict with you just by being different. In this sense, two gears meshing in a well-oiled, smoothly running machine are in conflict with one another. Notice, moreover, that the machine can operate only as a result of the various gears pushing against one another and forcing each other to move in ways they would not otherwise move.

What makes conflict so fundamental, I suggest, is that this *being different* is essential to my distinguishing myself from you and, in the same process, having my own identity. In a very powerful though subtle sense, I only *exist* by virtue of my resistance to you and to everything else around me. Except for these subtle conflicts, I couldn't even be distinguished as something separate.

I then typically consolidate these conflicts, and hence my identity, by making my version *better* from what I am in conflict with. I do not just have a different opinion than you, mine is truer than yours. Not only do I have a different religion, mine is the one God belongs to. I suggest that even when we don't verbalize or even recognize these views, they are present and fundamental to both *who* we are and *that* we are.

We have already had a glimpse of this process at the group level in the earlier discussion of in-groups and out-groups. There would have been no Protestantism except for the conflict with Roman Catholicism. You can't have Republicans without Democrats or labor without management. There can be no such thing as a group identity without that group conflicting, either grossly or subtly, with others. Ultimately, this reasoning devolves to tautology: you can't have difference without being different. Hence, I suggest, conflict is fundamental to our existence.

None of this is to deny our tendencies to come together, to join forces, to mimic each other, or to go along with the crowd. But notice that none of our harmonizing can take place except among individuals who are fundamentally separate, different, and, to that degree, in conflict with each other.

CHAPTER 10

THE GLOBAL PICTURE

Although many of the examples in this book are ethnocentri-
cally American, it is obvious that social structure is fundamental to all societies. Organized social life depends on shared
norms, values, systems of statuses and roles, and the like. As
we've seen, the forms social structure takes often differ from
one society to another. There are different languages, different
forms of government, different values—different ways of serving the same fundamental needs.

In this chapter, however, we'll see that there has been an
extensive homogenizing of the forms social structure takes
around the world. We are really becoming more and more
alike and part of one global family.

Go to just about any major airport in the world and you
may find it difficult to remember precisely where you are. You
could be in Los Angeles, San Juan, Paris, Honolulu, Auckland,
or Singapore and not be able to identify your location from a
casual view of the interior of the airport terminal. They are all
alike, although Singapore seems to work more efficiently than
most. Moreover, there is even a striking similarity among
smaller, less modern terminals: in small-island terminals such
as Kauai in Hawaii, St. Croix in the Virgin Islands, or Lombok
in Indonesia, for example.

In any of the major airports, the check-in counters look
pretty much the same. So do the rows of vinyl-and-steel seats
in the waiting rooms. Nowhere do you swing on vines or climb
a stepladder to get aboard your plane. No matter where you

are, you'll probably get aboard on an extending/contracting tunnel or a self-propelled staircase. If it's a large airplane, you'll probably find yourself boarding by row numbers—giving "families traveling with small children" and "passengers who need special assistance in boarding" a head start.

We could go on and on with this topic, but I think the point is clear. Despite very different histories and traditions in the various countries of the world, there are some striking commonalties in the social structures they now operate within. Since we are coming to share so much of the same social structure, it is all the more important that we understand how it works and how to control it. In this chapter, we'll look at some dimensions of what we share in common globally, putting most of our attention on some of our global problems.

World-Class Products

Years ago, international travel meant experiencing strange cultures that were always somewhat mystifying, often uncomfortable, and sometimes even distasteful. All three of these qualities could be applied to local food. While foreign travel still offers exciting new adventures, including the gastronomic, you no longer have to deprive yourself of your familiar favorites from back home.

In just about any major city, you will be able to eat at McDonalds, Wimpy's, Burger King, Wendy's, or KFC. You can relax with a Coke or Pepsi while watching a John Wayne Western on television. In 1992, Mickey Mouse and friends opened in a new Disneyland outside Paris amid great controversy from the French and others. Its success or failure was not immediately clear, but for our purposes, it is worth noting that no one asked: "What is Disneyland?" When you identify yourself as an American abroad today, you are more likely to be asked, "Have you been to Disneyland?" (Question: "Dr.

Babbie, you've just cured cancer, brought about world peace, and won three Nobel Prizes. What are you going to do now?" Reply: "I'm going to Disneyland!")

Realize that these comments about shared cultural artifacts do not only apply to American creations. Traveling at home or abroad, you may find yourself in the familiar seat of a Toyota (Japan), Honda (Japan), Volkswagen (Germany), Mercedes-Benz (Germany), Volvo (Sweden), Fiat (Italy), or Peugeot (France). Instead of a Coke or Pepsi, moreover, you can probably choose a Perrier (France), Heineken (Holland), Compari (Italy), Guinness (Ireland), or San Miguel (Philippines).

Years ago, anthropologist Ralph Linton[1] drew attention to the cross-cultural origins of those things we take for granted in our own culture. Part of his purpose was to poke fun at the anti-foreign-ideas fears of his time. Hence, he wrote of the "One Hundred Per Cent American."

> **There can be no question about the average American's Americanism or his desire to preserve this precious heritage at all costs. Nevertheless, some insidious foreign ideas have already wormed their way into his civilization without his realizing what was going on. Thus dawn finds the unsuspecting patriot garbed in pajamas, a garment of East Indian origin; and lying in a bed built on a pattern which originated in either Persia or Asia Minor. He is muffled to the ears in unAmerican materials: cotton, first domesticated in India; linen, domesticated in the Near East; wool from an animal native to Asia Minor; or silk whose uses were first discovered by the Chinese. All these substances have been transformed into cloth by methods invented in Southwestern Asia. If the weather is cold enough he may even be sleeping under an eiderdown quilt invented in Scandinavia.**

If you were to move into a hotel in Singapore today, you'd probably feel right at home. You might turn on your Sony television (Japan) and see former soccer superstar Pelé (Brazil)

---

1. Ralph Linton, "One Hundred Per Cent American," *The American Mercury*, 40 (1937), pp. 427–429.

doing an advertisement for Evian water (France). In order to remember your trip, you might load up your Nikon camera (Japan) with Kodak film (USA). We've gone several steps beyond what Linton described. Today we are all using the same brands.

World-Class Systems

Getting beneath the surface of globally familiar consumer products, we find widely shared systems for organizing and running human relations. Here are just a few that you will find in many, many places around the world.

Bureaucracy

In an earlier discussion, we examined bureaucracy as a system of social organization. It existed over three thousand years ago in China's Yin Dynasty, brought Europe out of feudalism, and is operating around the world today.

Some places, bureaucracies, often more mindless than even Americans are familiar with, were installed by colonial powers, most notably the British. Years ago, in a small Indian airport, I was instructed to complete a form in triplicate before I would be allowed to board my plane. Although the form made no sense to me, I complied. The clerk at the counter scrutinized the form carefully, separated the three copies, and then put each copy in a closet atop shoulder-high piles of presumably the same forms.

While stories of bureaucratic idiocies are entertaining and endless, we should not lose sight of the global commonalties in the benefits gained. In countries around the world, mail gets delivered, telephones ring, people pay taxes and get diplomas and marriage licenses, laws are passed, and rules are enforced. All sorts of activities occur in reasonably regular and rational patterns as a consequence of bureaucracy.

Suppose you were mugged in a foreign country. Where would you turn? To the police department? The American Embassy? The management of your hotel? Would you go to a bank for money to cover your loss? In short, you'd turn to bureaucracies. The bureaucratic infrastructure of societies around the world is so fundamental that we often lose sight of its existence.

Democracy

While there are plenty of nondemocratic governments around the world, the spread of democracy should be a source of amazement. When Seymour Martin Lipset[2] spoke of the United States as *The First New Nation,* he was recognizing America as the first in a series of radical experiments in self-government. Although democracy first appeared in the city-states of ancient Greece, with the philosophy of democracy resurfacing during the European Enlightenment, the creation of the United States of America as a democracy devoid of monarchy, aristocracy, theocracy, or military rule began a global revolution in the direction of self-government.

A little over two hundred years ago, democracy became an idea whose time had come. It has since spread from society to society around the world. In many cases, democratic political systems have been paid for in bloodshed. In others, ruling classes have gradually and peacefully relinquished their power. Moreover, democracy has gained nearly global agreement as the "right" form of government, such that unquestionably nondemocratic rulers feel obliged to adopt the trappings of democracy. Dictators like "Baby Doc" Duvalier, formerly "president for life" of Haiti, and Kim Il Sung, longtime president of North Korea, were periodically "reelected" by percentages in the high nineties. Today, no dictator will openly claim to rule for purposes of personal

---

2. Seymour Martin Lipset, *The First New Nation: The United States in Historical and Comparative Perspective,* New York: Basic Books, 1963.

gain; some claim must be made that the ruler is acting on behalf of "the people."

When the Kurdish rebels found themselves under siege by Iraqi forces in late 1991, freezing and hungry in desolate mountains, no one was shocked when the Kurds announced they would hold elections. The Iraqi Kurdistan Front said they were not attempting to establish an independent government; the new assembly would instead be "aimed at smoothing the decision-making process in the front and to give the elected leadership a popular mandate to decide on major issues affecting autonomy talks."[3] Democracy is widely accepted as an effective (though not necessarily efficient) form of governance in the modern world.

While democracy takes somewhat different forms, the similarities are striking. Many democracies separate executive, legislative, and judiciary functions. Bicameral legislatures with an upper and a lower house are common, and the executive branch typically is organized into major departments or ministries such as foreign affairs, defense, and agriculture.

Democratic principles, moreover, are not limited to formal government. It is not unusual to find democratic practices in labor unions, schools, voluntary associations, or boards of directors. None of this is to suggest that democracy is universal, nor that everything labeled democratic would live up to your image or experience. Nonetheless, it is a truly global system.

Economic Systems

For many people, the cold war was a dramatic, global struggle between capitalism and socialism as the dominant economic system for the world. To make a simple distinction, capitalism is based on private ownership of property and a free market for commerce, while socialism is based on state ownership of property and central control of commerce. Even

---

3. Associated Press, December 15, 1991.

before the collapse of the Soviet bloc, the battle was being won by a synthesis sometimes called "the mixed economy." Today, most countries have an economic system that is partly capitalist and partly socialist, with variations in leanings between the two poles and in official, ideological identifications.

Even the most ironclad of socialist economies have allowed for some private property or have at least tolerated some unofficial cottage industries, private gardens, or black markets. And fiercely capitalist societies allow for some state ownership, such as postal services and railways, and they all enact government regulations of industry and trade.

Labor unions are another aspect of modern economics common around the world. Beginning in the 1860s, British miners and textile workers discovered the power of banding together as a way of offsetting the considerable strength of the owners and managers. In some countries, such as Britain and Israel, organized labor now represents a major political party. In others, labor unions constitute a major constituency within a particular party, such as within the Democratic party in the United States.

More generally, collective bargaining and collective protests are firmly established techniques through which those who feel relatively powerless can gain power through the establishment of organizations, formal or informal. Consumer boycotts would be an example, and the phenomenon is also carried outside the economic realm, as in the case of student protests, for example.

The business corporation is another economic form found around the world. In contrast to owner-run businesses, the corporation separates ownership from management and often spreads ownership across a large number of shareholders. The professional management team—president, vice presidents, and so on—are overseen by a board of directors, who are charged with representing the interests of the absentee owners.

Recent years have seen the establishment of a special form of corporation: multinational conglomerates that operate in many industries and nations around the world. Gareth Morgan[4] identifies some of the best examples.

> **In 1982 there were 380 corporations with sales over two billion dollars per annum. The fifty largest corporations each had annual sales ranging from twelve to one hundred and eight billion dollars. Nineteen of these corporations had sales over twenty billion. The largest corporations, including names such as Exxon, Texaco, Mobil, Royal Dutch/Shell, General Motors, General Electric, Ford, IBM, Fiat, Unilever, and ITT, have annual sales figures that exceed the gross national products of many nations.**

There is a growing fear that such megacompanies often operate above the laws of any country. Their economic strength allows them to bring economic prosperity to local areas if not to whole nations, and they can just as quickly withdraw those benefits if they are not treated well by the host governments.

Without denying national differences altogether, it should be clear by now that certain economic structures and other social systems are common through most of the world. And the commonalties are growing steadily.

World-Class Figures

Just as there are brand-name products known around the world, there are also people whose faces and names are well known.

Politics

To begin, many political figures are easily recognized outside their own countries. The president of the United States, as well as the leaders of other major powers, are well known around

---

4. Gareth Morgan, p. 299.

the world. During the latter part of his presidency of the USSR, for example, Mikhail Gorbachev was more highly regarded outside his own country than within it. In an earlier era, Chairman Mao Tse Tung was one of the best-known political figures in the world. Some political leaders are well known for their misconduct, even when their world significance is relatively slight: Idi Amin of Uganda and Ferdinand Marcos of the Philippines are good examples from the recent past, joined more recently by Saddam Hussein of Iraq.

Some hereditary political figures, even those with little or no real authority, are quite visible on the global stage. The Queen of England and her family come to mind. People around the world saw Chuck and Di tie the knot and have watched the knot unravel.

Some political figures come to our attention as a consequence of focusing protest. Mahatma Gandhi and Martin Luther King, Jr., are models of this from our recent past. Nelson Mandela of the Union of South Africa and Lech Walesa of Poland are good current examples. Boris Yeltsin of Russia came to global attention in that fashion. Within the United States, Jesse Jackson is probably a good example of this.

Religion

The pope is a global religious figure, due to the number of Roman Catholics around the world. Billy Graham probably deserves a place in this category, due to his decades of worldwide evangelism. Mother Theresa is surely recognized around the world.

Other religious leaders become global figures primarily for their political activities. Examples would include Bishop Desmond Tutu of the Union of South Africa, the Dalai Lama (exiled from Tibet), and the late Ayatollah Khomeini of Iran.

Sports

Reporter Mitchell Stephens tells of the fighting in the mountains near Sarajevo during the spring of 1992. Serbian forces were firing down on the city. Stephens reports:

> **During a lull in the shelling, a Serbian soldier spotted an**
> **American reporter. Lowering his rifle, he drew the American**
> **aside and asked, "How are the Chicago Bulls doing in the NBA**
> **playoffs?"[5]**

Professional basketball, football, and baseball, centered in
the United States, are known around the world. Magic Johnson
and Michael Jordan are known to millions who will never set
foot in America.

Soccer is the most popular organized sport globally, and it
is generally conceded that Pelé, the superstar soccer player
from Brazil, was the best-known athlete on the planet during
his playing career. Muhammad Ali was undoubtedly better
known around the world than any other boxer in history.

Entertainment

Popular music has produced more than a few global figures:
Elvis Presley, the Beatles, Madonna, and Michael Jackson
come readily to mind. Sweden's singing group, Abba, is widely known. At his untimely death from cancer in 1981,
Jamaican reggae singer and composer Bob Marley was regarded by many as the first superstar from the Third World.

Jazz, that peculiarly American musical creation, is well
known worldwide, as are jazz legends such as Louis
Armstrong, Dizzy Gillespie, Miles Davis, and Charlie Parker.
Classical music, of course, has produced many world-class
performers. Enrico Caruso, Italian operatic tenor, was famous
early in this century—an amazing achievement, given the state
of mass communications at that time. Today, tenors Luciano
Pavarotti and Placido Domingo enjoy worldwide recognition.

Actors such as Marilyn Monroe, Greta Garbo, Sophia
Loren, Bruce Lee, Marcello Mastroianni, Toshiro Mifune, and
Sir Laurence Olivier are good examples of world-class figures
from the field of acting. World syndication of popular U.S.
television shows has created many other world reputations for

---

5. Mitchell Stephens, "Pop Goes the World," *Los Angeles Times Magazine,*
    January 17, 1993, p. 22.

actors, reflected, for example, in "Je t'adore Lucie," "Ich liebe Lucy," "Lucy ga suki" Ingmar Bergman, Federico Fellini, George Lucas, Steven Spielberg, and Akira Kurosawa are world-known directors.

The Media

Each of the preceding global figures owes his or her recognition in very large part to the global system of mass communications. In addition, the media have produced some global figures of their own. Edward R. Murrow, Walter Cronkite, Barbara Walters, and Ted Turner may have qualified for this position during their careers.

All these examples of global commonalties in consumer products, social systems, and personalities give evidence of the growing unity of culture shared by humans. As we'll see next, some of what we share is less enjoyable.

World-Class Problems

Many social problems can be found duplicated around the world, including crime, prejudice, and poverty. Some problems are not only found in most specific societies but are actually global in nature. During the cold war, the threat of thermonuclear war endangered everyone on the planet, not just those in the adversary nations. Hunger and overpopulation represent two other global problems, and I'm going to focus on them in the remainder of this chapter to illustrate the role of social structure in relation to such problems.

Hunger

On November 13, 1985, Colombia's Nevado del Ruiz volcano, dormant for nearly four hundred years, lit up the night skies with a violent eruption. In a single day an estimated 25,000 people were killed by the volcanic eruption itself or buried under the massive mudslides that resulted from it. The world was rightfully shocked by the tragedy and stunned by its magnitude. Newspapers around the globe headlined the news. Bogotá's *El Tiempo* suggested, "Perhaps God has forgotten us."

There was also a sense of powerlessness. Although the government was criticized for being too slow in warning people of the possibility of an eruption, and although better construction techniques and perhaps better public sanitation might have reduced the number of deaths, there remained the fact that little or nothing could have been done to prevent the eruption itself.

The Nevado del Ruiz tragedy illustrated the limits to our control over the natural environment, which is the context for all social life. When we must watch tens of thousands of people die in a single day—without being able to do anything to stop their deaths—we experience a special kind of impotence. As sociologists, moreover, there is little or nothing we can contribute to solving such problems.

Now consider another mammoth disaster. On November 13, 1985, the same day as the Colombian disaster, some 35,000 people around the world died as a consequence of hunger.[6] Three-fourths of them were children. Unlike the Nevado del Ruiz eruption and similar disasters, there were no newspaper headlines drawing our attention to the 35,000 who died of hunger that day. Quite possibly, you didn't know it happened.

There were no headlines on November 14, 1985, for the 35,000 people who died of hunger that day, nor for the 35,000 who died of hunger on November 15, 1985, nor for the 35,000 who died of hunger on November 16, 1985. There were no

---

6. It's worth noting that, strictly speaking, none of these people died of hunger per se. Rather, their chronically malnourished and undernourished conditions led them to die of such common ailments as the flu, measles, or diarrhea. If they hadn't been weakened by hunger, they would have survived those diseases, just as you and I do.

headlines, despite the fact that more people died of hunger during those four days than died in the atomic bombings of Hiroshima and Nagasaki combined.

Some 35,000 to 50,000 people die as a consequence of hunger every day of every month of every year. This amounts to some 13 to 18 million people every year. Not surprisingly, people have looked for ways to end this persistent global tragedy.

For the most part, people have looked for the answer to ending hunger within the field of agricultural technology. If only more food could be produced, surely that would end the problem. The "Green Revolution" of the 1960s was one result of this approach to the problem, and Dr. Norman Borlaug was awarded the 1970 Nobel Peace Prize for developing a high-yield strain of wheat that greatly increased the productivity of the planet. Yet hunger persisted.

There is by now a virtual consensus that more food is produced each year than would be needed to feed all the earth's people. One estimate suggests that the current production of grains and range-fed livestock alone could provide every human being with about the number of calories currently consumed by Americans. The persistence of hunger, then, is not simply a matter of too little food, and ending it is not a simple matter of producing more. When the National Academy of Sciences undertook a two-year, exhaustive study of the problem of world hunger, scientists concluded that we now possess everything necessary to end hunger, with one exception: *the political will to do it.*[7]

You'll recall that in Chapter 1, we looked briefly at the problem of war and noted how people have sought to bring about a lasting peace by building ever more powerful weapons. I suggested that the real answer to achieving peace over war lies in the domain of human social relations (i.e., the domain of

---

7. National Academy of Sciences, *Supporting Papers: World Food and Nutrition Study,* 5 vols., Washington, D.C., 1977.

Table 10.1
The Population Bomb

Year	Population	Doubling Time
0000	.25 billion	—
1650	.50 billion	1,650 years
1850	1.00 billion	200 years
1930	2.00 billion	80 years
1975	4.00 billion	45 years
1992	5.42 billion	41 years

Sources: Earl Babbie, *Sociology,* Belmont, Ca.:
Wadsworth Publishing Co., 1983, p. 561, and the
Population Reference Bureau, *World Population Data
Sheet,* Washington, D.C., 1992.

sociology). The same can be said of world hunger. More food is not sufficient to end hunger. An alteration in the human relations determining what happens to food is needed. Thus, where sociologists have little to contribute to the prevention of volcanic eruptions, we have a great deal to contribute to the elimination of world hunger.

War and hunger are not the only examples of major world problems that will depend on sociological attention for their solutions. Here's another that may be familiar to you.

Overpopulation

It can be argued persuasively that population growth around the world is the single most important cause of *all* environmental problems. There are simply too many people on the planet. With fewer people, our environmental problems would shrink, just as they exploded with the massive growth of population during the past two hundred years or so.

Table 10.1 lists some basic figures to describe what some have called "the population bomb." The nature of population growth is somewhat confusing, particularly with regard to where the primary problem lies. The sociological subfield of demography is addressed to the study of population and has provided a number of insights into this. Without attempting an exhaustive presentation of the subject, here are a few points that may be useful to you.

First, if you associate the problem of overpopulation with the poor countries of the globe, you are correct—in a sense. Approximately 80 percent of the current population growth occurs among those countries. Perhaps its most visible impact is in the form of hunger and death due to starvation. It is now estimated that 13 to 18 million people die of hunger and the disease it spawns each year, and three-fourths of them are children. Two centuries ago, Thomas Malthus observed that the growth in food production was being vastly surpassed by growth in the number of mouths to feed. In many countries, that discrepancy has played out tragically during our time.

Second, it is true that the developed nations have much lower growth rates today than are found in the Third World. In 1992, the growth rate for all of Western Europe was 0.2 percent; for Eastern Africa, it was 3.2 percent. Whereas North America was growing at a rate of 0.8 percent, Central America, immediately to the South, had a growth rate of 2.5 percent.[8]

As they have sought to understand such differences, demographers have uncovered an important societal pattern. They refer to this discovery as the *Theory of Demographic Transition.*

The Theory of Demographic Transition. To understand this theory and its relation to population growth, let's begin with the concept of *zero population growth* (ZPG), the condition in which the size of a society's human population would remain stable over time. In the simplest analysis, it would seem that ZPG would be achieved if every woman had two children, or if the average for all women was two children. Since about half

---

8. The Population Reference Bureau, *World Population Data Sheet,* Washington, D.C.: Population Reference Bureau, 1992.

of those children would be female, maintaining the pattern of two children per woman would result in no growth in numbers.

To be a bit more precise, however, a pattern of two children per woman would actually result in a *decline* in population, since some of the females would die prior to producing their two children. Thus, in a society such as the United States, with a relatively low level of infant mortality, an average of 2.17 children per woman would produce a pattern of population stabilization over time. Low rates of infant mortality are a recent development in human history, however.

At the outset, women gave birth to large numbers of children—eight or ten was not unusual—and most of those would die before reaching their own childbearing years. The matching of very high birth rates with very high death rates kept the human race alive and relatively stable for about three million years. In some of the poorest countries of the world today, one child in five will die prior to its first birthday. This is contrasted with less than one in a hundred in some of the wealthier countries.

Recently, however, death rates around the world have been reduced dramatically. This has happened as a result of increased food production, improved health care and public sanitation, and other factors. Wherever this has happened, a massive increase in population has occurred, since birth rates have stayed at the levels previously required for survival. Women continue giving birth to large numbers of children, but many of those children survive to adulthood and produce large numbers of children themselves.

This pattern, however, is accompanied by increased urbanization of societies and typically by industrialization. The final stage in the historical pattern described by the theory of demographic transition is one in which urbanization and industrialization result eventually in reduced birth rates.

Seven countries serve to illustrate the three stages of the demographic transition as of 1988. Sierra Leone is a small country in West Africa with high birth and death rates and an overall growth rate at about the world average of 1.7 percent a

Table 10.2
Three Stages of Demographic Transition

	Annual Growth Rate	Crude Birth Rate	Crude Death Rate	Percent of Pop. Urban	GNP Per Capita
Sierra Leone	1.8%	47	29	28%	$ 310
Kenya	4.1%	54	13	19%	$ 300
Zambia	3.7%	50	13	43%	$ 300
Nicaragua	3.5%	43	8	57%	$ 790
Japan	0.5%	11	6	77%	$12,850
Belgium	0.1%	12	11	95%	$ 9,230
West Germany	-0.1%	10	11	94%	$12,080

Source: The Population Reference Bureau, *World Population Data Sheet,*
Washington, D.C.: Population Reference Bureau, 1988.

year. Kenya, Zambia, and Nicaragua, as you can see in Table
10.2, have lowered their death rates dramatically (compared to
Sierra Leone) but their birth rates have remained very high,
resulting in high annual growth rates. Japan, Belgium, and
West Germany (these data are prior to unification) are examples of countries that have lowered their birth rates to match
their death rates, in the vicinity of population stabilization.

Eighty percent of the current world population growth is
occurring in countries where death rates have begun to drop
dramatically but birth rates remain very high, in societies still
in the process of urbanizing and industrializing. This may suggest to you that the problem of overpopulation can be solved
eventually by letting the demographic transition run its course.
This leads to the third point I want to make regarding overpopulation.

While the bulk of population growth is occurring in the
poorer countries of the world, there is good reason to say that
it is precisely the urbanized and industrialized nations that constitute the real problem of overpopulation. It is the rich coun-

tries that consume most of the world's resources and generate most of its pollution. This issue served as a main point of controversy during the 1992 "Earth Summit" in Rio de Janeiro. The United States was one of those urging less-developed countries to stop cutting down their rain forests—which adversely affects the global weather, among other things. When the developed countries were asked to curtail activities that produced carbon dioxide—the primary cause of global warming—the United States stood alone in insisting that the agreement not specify any targets or timetables for doing so. The United States is far and away the chief producer of carbon dioxide.

In summary, overpopulation presents a serious danger to the well-being of the planet and its inhabitants. This is true of overpopulation among the poor nations and it is true among rich nations as well. We are simply producing more babies than the planet can accommodate, and it is likely to get very ugly as we approach the absolute limit of population that earth will tolerate.

There's no denying that there is some point beyond which humans cannot continue increasing their numbers. As a mathematical exercise, demographer Angsley Coale once calculated that if we continued our present growth rate until the year 3000, the mass of human flesh would be expanding outward from the earth's surface at a rate faster than the speed of light.[9] He wasn't predicting that would happen, of course, because we would have run out of food and water long before that could happen. The question is: how much before that should we or must we stabilize our numbers?

Most people can see some evidence of problems now, even if they don't always identify the problem as overpopulation. Impoverished people starve or watch their children die of diseases like measles and diarrhea. People in rich countries can

---

9. Cited in Charles Nam, ed., *Population and Society,* Boston: Houghton Mifflin, 1968, p. 63.

see the disadvantages of crowding as they stand in line at stores, sit in gridlocked traffic, or get closed out of courses they want at the state university.

So why do people keep producing more babies than we need? It might be useful to consider what might be necessary in order to stabilize world population. First, for example, we would need to know what causes babies to happen. Presumably, people didn't always know that, but we got the answer some time back. Still, people keep producing more babies than are necessary for replacement.

Self-Interest and Childbearing. Part of the reason is that many babies have been born to people who didn't want them, but who were unwilling to give up the activities that cause babies. Under those conditions, it would be necessary to learn how to have sex without having babies. As you know, we learned the answer to that as well.

Despite having the ability to pretty much choose the number of children we want, the human race still produces far too many babies. There must be some other reason why that's so: something that has nothing to do with glandular biology, reproductive chemistry, or the physics of latex.

Let's consider two sociological factors that help to explain the continued overproduction of babies. The first set of factors is properly called *social psychological*.

Some people *want* to have more babies than are necessary for replacement. And some who might not want a lot of children fail to take steps to avoid it. The most logical explanation for this would be that having a lot of children may be of benefit to the parents. One of the central themes of this book and of sociology in general is that individuals often act in their own self-interest even when it is damaging to society as a whole.

In earlier times, there were a number of self-interest reasons for a couple to have many children. Before governmental and employer retirement programs, children were your main source of security in your old age. Similarly, when farming is labor-intensive, there is value in having a small army of workers you don't have to pay.

In a modern, industrial society, however, the advantages of large families have disappeared for almost everyone. In fact, as we have extended the period of dependency of our children—with required schooling, for example—children have become a poorer and poorer "investment" in terms of pure self-interest. The U.S. Department of Agriculture has recently estimated that it costs an average of $100,000 to raise a child to age 18.[10] Moreover, couples are more likely to have children at a time when they are earning entry-level incomes and could use the money they are spending on their babies. Add to this the disruption of the mother's ability to work, and it becomes clear that having babies, especially a large number, is a heavy financial burden for a modern American family.

While I've mentioned couples having babies in the preceding discussion, one-fourth of all babies born in the United States in 1988 were born to unmarried mothers, typically representing a difficult financial situation for the mother and child. The heaviest burden falls upon young, unmarried girls, who may have to drop out of high school and find ways of supporting their children with little or no possibility of earning a substantial income. In 1988, one baby in twelve (a total of 322,400 babies) was born to an unmarried teenager in America.[11]

Part of the reason for the excess of babies is that they "just happen." In 1992, the World Health Organization (WHO) estimated that 100 million sex acts around the globe every day resulted in some 910,000 conceptions. Half of the conceptions, moreover, were unplanned, according to the WHO estimates.[12]

To understand why young people would have *any* children, let alone many, it is necessary to look beyond self-interest. The

---

10. Blayne Cutler, "Rock-a-Buy Baby," *American Demographics,* Vol. 12, No. 1, January 1990, p. 38.

11. U.S. Bureau of the Census, *Statistical Abstract of the United States,* Washington, D.C.: U.S. Government Printing Office, 1991, p. 67.

12. Associated Press, June 24, 1992.

answer lies in the realm of *social structure*. This is the second set of factors, and we'll consider two aspects: (1) beliefs and values, and (2) pro-natalist organizational structures.

Beliefs and Values. In a great many ways, we are all socialized to the view that having children is a good thing. If you felt uncomfortable during any of the preceding discussions about the problem of overpopulation in general or the costs of having children in particular, that discomfort is evidence of your own socialization to this view. Perhaps American politics can serve as a bellwether for this value. We speak of politicians kissing babies as evidence of their commitment to *family values,* another political buzzword.

Liberals are quick to leap to the defense of children as our "most important resource for the future" whenever conservatives threaten to cut school funds; conservatives are likely to defend the "rights of the unborn" whenever liberals suggest women should have the right to choose abortion.

In contrast to all this, imagine a politician who built a campaign around the view that children were a drain on the economy, and families should be taxed higher for having them. Every new crop of babies means a need for school buildings and teachers, for example, both of which cost the taxpayers money. In the longer run, they represent increased pollution, crowding, and digging up farmland for more houses and strip malls. Besides all this, they're often noisy and have a tendency to drool when they are very young and break things when they are older. I think you can imagine the powerful set of arguments that could be marshaled for this platform. Still, it doesn't seem likely that any political candidate would be elected on it. Babies have too powerful a lobby, and most of us are members of it.

A set of established beliefs also drives young people to have babies, even when it is not in their interest to do so. Consider just a few; see which ones you share.

◆ A man isn't *really* a man until he has made a baby, and the more he has, the more virile he is.

- A woman isn't *really* a woman until she has had a baby, and the more she has, the more loving and generous she is.
- You owe it to your parents to provide them with grandchildren.
- God wants you to have children.
- Having children will hold a marriage together.
- If a man doesn't perpetuate his family name, his ancestors will be disappointed in him.

These are just a few of the values and beliefs that motivate people to have baby after baby, loud, drooling, and destructive though they are. These are some of the reasons you may have found some of the preceding comments "in poor taste." Social structure is real and has a powerful impact on us.

Pro-Natalist Organizational Structure. Most societies have rules and regulations that encourage people to have children, even though they may not be justified that way. Let's look at a few of the pro-natalist features of American society.

If you pay federal income taxes, you are given a personal deduction worth $2,300 per dependent for 1992 and increasing each year. The more children you have, all else being equal, the less you pay in taxes. As we just saw, each new child will cost the society more in tax-supported facilities, so this tax break makes no sense—except as a social commitment to producing as many babies as possible.

Every now and then some politician suggests that the dependents' deduction be abolished, and the objections are quick and loud. It is argued, in part, that this represents the only deduction most poor people can take. This is, of course, an ironic justification in view of the financial burden children represent to a family—far in excess of the value of the deduction. This is something like giving free heroin to the poor, since they can't afford vacations abroad. The "gift" leaves the recipient worse off than before. If the aim were to benefit the poor, a more logical solution would be to raise the income threshold at which people start to pay taxes, rather than giving everyone—rich or poor—benefits for producing children.

The discussion of pro-natalist policies in American society could be extended at length. Resistance to sex education in the public schools, for example, has a consequence of increasing unwanted pregnancies, as does resistance to giving students condoms. Legislation that reduces a woman's freedom to elect an abortion instead of childbirth increases the birth rate. When the Bush administration prohibited federally funded physicians from mentioning the option of abortion, that increased the birth rate—especially among poor women. This is not to suggest that any of the policy makers responsible for these actions *intended* to increase birth rates, but the policies were pro-natalist all the same.

The pro-natalist implications of policies are not always that obvious. Consider the matter of traffic. It will probably not surprise you to learn that a 1992 survey of the nation's highways found 40 percent were exceeding their design capacities.[13] Clearly, population growth contributes directly to congested traffic: more people, more traffic.

One of the solutions to highway congestion has been the designation of carpool lanes, reserved for vehicles with some specified number of passengers such as two or more. Sometimes, carpoolers are given free passage through commute-hour toll stations. The idea has been that unrelated individuals would join up and travel together in order to get the various benefits. As an unintended consequence, the special break for "carpools" has been to benefit large families, since they are the most likely to have more than one family member traveling at any one time. Thus, ironically, those most responsible for creating the future traffic congestion have been rewarded by special assistance with today's congestion.

Can Anything Be Done? I know I've pursued this issue at some length. The depth of the pro-natalism that I've been describing may very well have made you somewhat uncomfortable about this discussion. It's just not nice to seem to say

---

13. Associated Press, July 13, 1992.

bad things about babies and little children. If you have experienced any discomfort, I hope you will use that personal experience as a sociological observation about how social structure affects us on this topic.

For all the pro-natalist elements in our social structure, it is nonetheless possible to deal with the issue of overpopulation. It is possible to overcome all the social structural support for large families, as demonstrated by the People's Republic of China: the world's largest nation with roughly a fifth of the globe's population.

Despite a very long tradition of reverence for the family and support for large numbers of children, the PRC explicitly reversed this value. The traditional, Confucian ideal was to have four generations living under the same roof: young children with their parents, grandparents, and great-grandparents. Thus, it would not be unusual to find twenty or thirty family members living together in a single household.

With the Communist Revolution in 1949, the government began systematically undermining the authority of the extended family in several ways. Large family landholdings were broken up, as part of the overall economic reforms. Arranged marriages were forbidden, again reducing the power of the family. Students were sent to colleges far from home. And at all levels of schooling, young people were resocialized to the view that loyalty to the Communist party took priority over loyalty to the family.

Beginning in 1971, the Chinese government pressed a national policy of *wan xi shao* (later–longer–fewer), to encourage people to (1) delay childbearing to later in life, (2) have a longer space between children, and (3) have fewer children. Rural families were encouraged to limit their families to three children, with urban families stopping at two. By 1977, the limit was two children for all couples; two years later it dropped to one child per couple.[14]

14. Susan Greenhalgh, "Socialism and Fertility in China," *The Annals of the American Academy of Political and Social Science,* July 1990, Vol. 510, pp. 73–86.

Table 10.3
Birth Control in People's Republic of China (millions)

	IUDs Implanted	Vasectomies	Women Sterilized	Abortions
1981	10.34	0.65	1.56	8.70
1982	14.07	1.23	3.93	12.43
1983	17.76	4.36	16.40	14.37

Source: Robert Delfs, "The Fertility Factor," *Far Eastern Economic Review,* July 19, 1990, p. 19.

Couples who refused to cooperate with the national policy were sometimes fined, sometimes punished in other ways. In 1982, the Central Party Document Number 11 of 1982 directed local leaders to regain control of family planning. Various forms of birth control were greatly increased, as Table 10.3 indicates.

Despite the difficulty of enforcing national family planning policies in such a large country, particularly in the remote rural areas, China's population growth was noticeably affected. Susan Greenhalgh reports that the crude birth rate dropped from 33.43 to 17.82 between 1970 and 1979. Even though the death rate declined from 7.60 to 6.21 in the same years, the rate of natural increase was cut by more than half.[15]

Clearly, it is possible to change the pro-natalist elements of social structure. While you may not support the methods used by the PRC, the point of this example is to indicate that fundamental changes can be made to social structure at all levels. We created the social structure that governs us; we can change it. At the same time, you should not imagine that it is a simple matter to do so, even when logic and both individual and collective self-interest dictate it.

---

15. Susan Greenhalgh, July 1990, p. 75.

As we've seen throughout this book, social structure is designed for survival: its own survival. Taking it apart and redirecting it is a delicate task.

A Sociological Paradox

I will close this chapter and the book on a very different, almost contradictory, note. If massive social change is possible, where and how does it begin? What's the impetus for a complex social system to turn a corner and begin lumbering mindlessly off in a different direction? The answer, as you may have guessed, is ultimately the individual. The late Buckminster Fuller was fond of asking, "What can the little individual do?" What can one person, having no official position or great wealth, for example, do to change the direction of massive and complex social institutions? Bucky's answer was that *only* the individual could have that impact.

The systems we create cannot change themselves, except in the predictable ways we've designed them to change. Only people can initiate those changes, and frankly, it always starts with one person. There's a bit of a paradox here. Very little is ever accomplished by only one person, and yet nothing ever happens without one person stepping out of line and taking responsibility for the change—as though he or she were going to do it alone.

The question we always face is "Who will be the one?" and the obstacle is that we always think someone else is better suited to the task. Even if we think we could do it, we worry about what others will think of us. Will they think we're on an ego trip or being self-righteous? Maybe we'll look naive, goody-goody, or just plain stupid. What if we fail or make matters worse? The reasons for doing nothing are persuasive.

One of the most joyful experiences I've had in recent years involved my research for a book about individuals taking personal responsibility for public problems.[16] I was often deeply moved by the powerful examples of ordinary individuals stepping forward to do what needed to be done.

On December 1, 1955, Rosa Parks was a seamstress in Montgomery, Alabama, riding the public bus home after a hard day's work. In accord with the established norms of that time and place, Ms. Parks was riding in that section in the back of the bus set aside for black people.

As the bus became more crowded in the rush-hour commute, however, the white section of the bus filled up, and the driver called back for those seated in the first rows of the black section to stand up and let the white passengers sit down. Today, it's hard to appreciate how absolutely those black passengers had no choice in the matter, given the norms of the times. They might not have liked it, some might have even grumbled a little under their breath, but it was simply their cross to bear.

It might be useful for you to imagine yourself arriving at the post office today and finding a long line with only one clerk on duty. Consider further that (1) you could take your place in line, maybe grumbling a little, (2) you could leave the post office rather than put up with the long wait, or (3) you could rush forward, leap over the counter, grab the postmaster by the lapels, and demand that more clerks be put to work on the counter.

I don't suppose you can seriously imagine yourself taking the third alternative, and I'm not suggesting you do that. It would be outrageous for you to take matters into your own hands that way. At the very least you'd get arrested, and it's hard to predict how you would ultimately end up, because that kind of behavior is hardly sane. You couldn't even count on your fellow sufferers to support you, because your irrational rebelliousness would interrupt even the slow service that was being provided.

---

16. Earl Babbie, *You Can Make a Difference*, New York: St. Martin's Press, 1985.

If you take the third alternative seriously, however, you may begin to get some sense of the situation Rosa Parks faced that winter day in Montgomery. She could have gotten off the bus and walked home rather than being pushed around, but what would that have proved, really? All things considered, the most sensible course of action was that chosen by her fellow black passengers: get up, move to the back of the bus, and hope you got home before much longer.

Rosa Parks said no. Even though she didn't rush to the front of the bus and grab the bus driver, her simple refusal to give up her seat was every bit as violent an assault on the established system: it was an outrageous attack on the way things had always been, the way things were supposed to be, the only way things could function for everyone. It was an assault on every element of American society that was intertwined with segregation in the South. All that notwithstanding, one tired black lady said no.

As you probably know, Ms. Parks was arrested and thrown in jail for her act of defiance. That probably should have been the end of it. She could have served some time in jail, paid a fine perhaps, and she would have been an example to all other black Americans to stay in their place.

A number of the local black clergy, however, felt that Ms. Parks had been treated so inappropriately that some kind of protest needed to be made. Someone had the idea of a one-day symbolic boycott of the bus system. If a sizable proportion of the black community gave up riding the buses for one day, that might have some impact. It might at least keep things from getting worse.

Even a one-day, symbolic boycott would require a great deal of organization, however, and whoever took on the job would be running the risk of retaliation from the white community. As they looked among themselves for an organizer, the clergy asked Martin Luther King, Jr., to take on the job. Dr. King probably had as many reasons for saying no as Ms. Parks had for saying yes. He was a new minister in town, just starting his ministerial career. In effect, he was being asked to risk his career and perhaps his physical well-being for what could be little more than a hollow gesture. And he said yes.

To everyone's surprise, the black boycott of the Montgomery bus system was virtually total. People formed car pools, they walked to work, some rode bicycles. The experience of oppression was pushed aside by one of pride and dignity. The boycott was extended. Soon the police began harassing black carpoolers, pedestrians, and bike-riders. Dr. King and others were beaten and thrown in jail. And the boycott continued.

Soon, blacks in Montgomery and across the South had captured the attention, admiration, and support of people around the world. In November 1956, less than a year after Ms. Parks's lonely act of conscience, the U.S. Supreme Court declared that racial segregation in public facilities was unconstitutional.

By their willingness to step out of line, Rosa Parks, Martin Luther King, Jr., and thousands more "little individuals" brought about a revolution in race relations in America and around the world. The same blind, lumbering juggernaut of social institutions that made racial segregation inevitable was turned enough in its tracks that it now lumbers on, sometimes slowly and sometimes blindly, toward an unfolding racial equality. Only an individual can make that happen. Although we can easily mislead ourselves by thinking that each of us has the same power to affect changes in society, we can also mislead ourselves by thinking we don't.

You are left, then, with something of a paradox. Social institutions have the power to grind up individuals by the thousands and millions. Moreover, the institutions we create seem to take on lives of their own, ensuring their own survival, even at the expense of ours. Clearly, no individual is a match for a complex social institution. Yet any challenge to an institution must begin with an individual. Ultimately, institutions don't act: neither do organizations or groups. Only the individuals involved can decide and act, and in any instance someone must take the first step. At the same time, individuals almost never make decisions or take actions in isolation. They are always influenced to some degree by those around them and by the nature of the social structures they live in.

Society, then, is a paradoxical interplay between our individual and corporate identities and lives. Sociology is devoted to the mastery of that paradox.

APPENDIX

SOME POINTS OF VIEW ON SOCIOLOGY

In the following pages I have reprinted a copy of an article that originally appeared in the journal *Teaching Sociology*. It was my attempt to summarize the core of the discipline as briefly as possible. I have also reprinted the comments of four other sociologists who were asked to evaluate my summary: James A. Davis, Michael R. Leming, Laura Kramer, and John J. Macionis.

THE ESSENTIAL WISDOM OF SOCIOLOGY*

Earl Babbie

Chapman University

Although the introductory sociology course typically conveys some information about society (primarily American society), I think most sociology instructors would agree that it is more important for students to grasp some fundamental sociological *concepts:* role, interaction, structure, conflict, socialization, and so on. I am not so naive, however, as to imagine that we would ever agree on precisely *what* the most important concepts are.

In my own teaching and writing, I have found it even more important to ground students in some basic principles or *meta*-concepts that distinguish the sociological view of things and that make sense of the concepts we weave together in the sociological construction of society. Below I will identify and describe briefly ten basic principles, not in the hope of reaching agreement with my colleagues, but with the idea that our disagreement might be productive.

1. Society has a *sui generis* existence and reality.

We are indebted to Emile Durkheim for the assertion that society cannot be reduced to its presumably component parts. You can't understand society fully by understanding only the individual human beings who constitute it.[1] Here are some illustrative examples:

Earl Babbie, "The Essential Wisdom of Sociology, *Teaching Sociology,* October 1990, 526–530. Revision of a paper delivered at the annual meetings of the American Sociological Association, San Francisco, August 13, 1989.

1. Indeed, it may be that the "individual human being" is not the element of which societies are composed. A strong case can be made that the critical element is *social status*—which then serves as the point of intersection between individuals and society.

Few people want war, but wars occur all the time.
A substantial majority of the American public wants gun control, but we don't have it.

This point has numerous implications. For example, it can redirect efforts to identify and deal with social problems, allowing for the possibility that some problems stem from the way in which society is structured and functions rather than from evildoers. The fact that women continue to earn only about 60 percent as much as men in the United States, the fact that black American babies have twice the infant mortality rate of white American babies, and similar social facts cannot be explained in terms of individual human beings doing bad things.

Recently this matter was illustrated dramatically when the U.S. Supreme Court refused to declare capital punishment racially discriminatory on the basis of incontrovertible statistical evidence. Instead, they said, it would be necessary to show that individual jurors had made their decisions on the basis of racial prejudice. In other words, the Court was unwilling to consider the possibility that a *system* could be racist, whether or not the individual participants were racist.

2. It is possible to study society scientifically.

In the fine tradition of Auguste Comte, we must point out to students that knowledge about society can be more than learned beliefs or "common sense." It is possible to study society scientifically, just as we study aspects of the physical world.

I find the potency of probability sampling an effective vehicle for making this point. Students are impressed to learn that a sample of 1,500 voters can predict so accurately the eventual votes of tens of millions. I remind them that social scientists sometimes are accused of taking all the suspense out of elections—because we are so accurate.

Some students are convinced by the classic examples of Stouffer's (1949, 1950) work on relative deprivation. After presenting the seemingly trivial "findings" that (1) black soldiers had higher morale in northern than in southern training camps, (2) soldiers in the Army Air Corps were the most likely to think the promotion system was fair, and (3) the most highly educated soldiers were the most likely to resent being drafter, I reveal that Stouffer discovered just the opposite of each of these "findings." Then we set about understanding how the concept of relative deprivation makes sense of such puzzling discoveries, as follows:

(1) Black soldiers in southern training camps considered them-
    selves better off than the black civilians around them; black
    soldiers in the north, however, were not doing as well as
    black civilians in the north.
(2) Soldiers in the Army Air Corps, where promotions general-
    ly were rapid, were likely to know of someone who had
    been unfairly promoted faster than themselves; in contrast,
    there was little opportunity for apparent unfairness in the
    Military Police, where promotions were slow for everyone.
(3) Less highly educated soldiers were more likely to have
    friends working on assembly lines or as farm laborers—
    occupations that were exempted from the draft—whereas
    more highly educated soldiers were unlikely to have friends
    who were exempted. Thus the less highly educated soldiers
    were more likely to feel that they had been treated unfairly
    in comparison with their friends, whereas the better-educat-
    ed soldiers felt that they had been treated the same as
    everyone else.

Sociology is the perfect vehicle for training students in crit-ical thinking, and we provide that training in demonstrating the use of logic and empirical observation as tools of rigorous dis-covery.

3. Autopoiesis: Society creates itself.

I know that the term, *autopoiesis,* which I take from the Chilean biologist/philosopher, Humberto Maturana (1980), is not well known to sociologists. Moreover, Maturana has said that the term does not apply to society—but then, he's not a sociologist.

Autopoiesis might be defined as "self-creation." It is easiest to understand the term through linguistic examples. Some statements, for example, become true in their assertion. That is, saying them makes them true, as in these illustrations:

I apologize.
I am speaking.

I think that sociology can offer the powerful statement that society is autopoietic: society creates itself. If society can be seen as a network of norms, sanctions, values, and beliefs, for example, where did all those elements come from? Most of us have long since given up the notion that they came from gods or from natural law. The norms that govern the operation of society are generated in the operation of society.

The importance of this view for students is that it undercuts notions about what is "really" so. Monogamy, democracy, honesty, politeness, and consideration for others may be norms of modern American society, but they also were *produced* by American society.[2] This concept, if grasped, provides a fertile ground for the establishment of cultural relativity in place of ethnocentrism.

4. Cultures differ widely across time and space.

Here we join our anthropological colleagues in pointing out that all of creation is not white, American, and middle-class. We understand this point, but it is a shocking eye-opener to many freshmen.

Gaining awareness that differences exist, however, is only the beginning. Our second task in this regard is to undermine students' implicit ethnocentrism, offering the possibility of tolerance and even of cultural relativism as an alternative approach to the world. Insofar as students can begin to see cultural forms as merely different rather than better or worse, we have offered them a point of view that will serve them and those with whom they share the planet.

---

2. This is not to say that they are unique to us. They also were self-created by other societies.

I do not mean by these comments to urge a dogmatic egalitarianism as regards cultural differences. I would hope that students emerge from their studies favoring democracy over totalitarianism, racial tolerance over prejudice, and any number of other value-based positions. Our aim, however, should be that such choices are made on the basis of analysis and reflection rather than simply by default through cultural inheritance, perpetuated by ignorance.

5. The individual and the society are inseparable.

Earlier I noted that society cannot be reduced to the level of individuals: society has a *sui generis* existence at a different level. Now, I wish to verge on suggesting that individuals are merely figments of society. Without going quite that far, however, I want to point to another contribution that sociology can make to students.

Individual identity is strongly *sociogenetic*. Often I engage students in a conversation about "who they are." Typically I begin with the Kuhn-McPartland (1954) Twenty-Statement Test, in which I ask them to write out 20 answers to the question, "Who am I?" We spend some time finding patterns in their responses; eventually we reach the realization that virtually all the answers they offer—student, daughter, baseball player, liberal, and so on—are *social statuses*.

Although we also spend time on the question of who they *really* are—and discover that we can't express the answer satisfactorily in words—they grasp the point that they express their identities totally in terms of their standing in society. If there were no society, none of their "identities" would apply.

The advantage of this realization, I think, is that students eventually can come to see how they occupy social statuses, and that they could choose to occupy different statuses if they wished to do so. Moreover, this distancing from their statuses makes them less vulnerable to attack by others. When someone complains that students are noisy and irresponsible, my students are less likely to take that complaint as a personal attack and can ask instead whether it's true of them.

6. Systems have system needs.

Along with the discovery that society is an entity, students also can gain from sociology the important insight that society—as a system—has "needs." This message carries the risk that students may tend to anthropomorphize society, assuming that it has desires, motives, and similar human qualities. As a remedy, it's useful at times to point out that an automobile also has needs, though these needs—gasoline, oil, electricity, cooling— are very different from those of a human being. I find that this analogy helps students see that a system can have needs which reflect its particular structure and operation.

Rather than attempting to walk students through Parsons's "AGIL" or any similar system, I usually discuss replacement as a fundamental need for a society. People die, so some provision must be made for the production of replacements. Initially this point involves a discussion of the various forms of the family, but it goes beyond that topic to socialization, because an infant is hardly a replacement for a rocket scientist (or even for a sociology professor, for that matter).

Usually, I spice things up with a brilliant but largely unknown book by John Gall (1975), entitled *Systemantics: How Systems Work and Especially How They Fail.*[3] Gall discusses how systems tend to take over, once they've been created. One of his key points is that the survival interests of the system often gain precedence over and work against the purposes that led to the creation of the system in the first place.

An appreciation of system imperatives is important to students, I think, when they set out to change social conditions that they find unacceptable. If the processes creating those conditions are linked to system survival needs, students should be aware of that link, or else their efforts for change will be ineffective.

---

3. I suspect that this book has not been taken seriously by sociologists because—like *Parkinson's Law* and *The Peter Principle*, two books that it emulates—it is too much fun. Somehow we assume that important analyses must be boring.

7. Institutions are inherently conservative.

To follow from Point 6, it is important for students to realize that the first function of any institution is survival. If there were ever institutions whose purpose was self-eradication, they no longer exist to educate us about the genre. Rather, institutions are structured so as to ensure their own survival. They are fundamentally conservative in the sense of "conservation," not in the conservative-versus-liberal sense.

Moreover, in an integrated society, institutions support each other's survival. The government pays for schools that tell students to be law-abiding; parents take their children to church, where they are told to revere and obey their parents. Yet the inherent conservatism of institutions as institutions does not prevent them from serving also as sources of change within the society as a whole. This situation is likely, for example, when the interests of different institutions conflict— such as those of religion and education.

The idea of institutional conservatism usually provides a fruitful context in which to introduce other basic notions: socialization, internalization, and the ways in which individual identity is woven into institutional structures.

8. Explanatory sociology is implicitly deterministic.

This topic is often troublesome for students. I suggest that *explanatory* social science is grounded implicitly in a deterministic image of human beings. When we set out to discover the *causes* of prejudice, for example, we assume that prejudice is *caused* by something and that free choice is not one of the possibilities. In somewhat different terms, we operate with a model that assumes that human behavior is determined by forces, factors, and circumstances that the individual actors are unaware of and/or cannot control. Put more crudely, this model suggests that human beings—including you and me— do not have the free will we imagine we have.

I've discussed this issue at length elsewhere (Babbie 1986, 1989), so I'll avoid a recapitulation here. Sociologists do not *believe* that humans lack free will, but the model we use for explanation assumes that to be the case. I point out to students that whether we have freedom in any ultimate sense, all of us tend to live our lives in largely deterministic ways. It is useful to point out the extent to which we give away our freedom in our language, as when we say: "You make me so angry," "The IRS won't let us do that," or "I could never talk to him after he said that."

9. Paradigms shape what we see and how we understand.

It is important to communicate the idea of paradigms to all students, and sociology is the perfect vehicle for doing so. Students need to learn that paradigms are ways of looking at life, but are not life itself. They focus attention so as to reveal things, but inevitably they conceal other things, perhaps rather like microscopes or telescopes. Paradigms allow us to see things that otherwise would be hidden from us, but they do so at a cost.

Paradigms also color our vision, as in the analogy of colored glasses. If you wear red-tinted glasses, everything looks reddish; blue glasses give the world a bluish tinge; and so forth. Students can see that point, and the linchpin of the analogy is the assumption that you have to wear some set of glasses in order to see. This point helps students to appreciate the dilemma of whether we can ever know what's "real" or "true."

Of course, it is important to familiarize students with the three basic sociological paradigms: structural functionalism (or social systems theory), interactionism, and conflict theory. If they seem able to grasp these broad distinctions, you can add such paradigms as role theory, exchange theory, and labeling theory. The key point is that each of the sociological paradigms offers powerful insights into the nature of social life, but that none of them represents the whole truth.

You can make an even more powerful contribution by helping students to see *rationality* as a paradigm. At the outset, students will regard rationality essentially as "the way things are" or at least as "the way they should be." Whenever I ask students to think of alternatives to rationality, they seldom go beyond "irrationality." Usually they get a little nervous when I seem to suggest that rationality isn't the "true" or even the "best" paradigm.

Rationality may have given us Teflon, but it isn't much good for the appreciation of poetry. Further, it has little to contribute to the experience of love or of making love. In a moment of passion, for example, when your partner shouts, "God! I love it!" don't ask how much or point out that you are not actually a god.

Rational sociological paradigms aren't much use at such a moment, either. Don't go functionalist and ask, "Do you want to know how this works?" Don't put on your interactionist cap and ask, "What exactly does this symbolize?" And no matter how strongly you may be committed to a conflict perspective, never say, "Did you know that this is what capitalism does to the working class?"

10. Sociology is an idea whose time has come.

I'd like to conclude these comments on a possibly chauvinistic note: all of the major problems that face us as a society and as a world are located within the territory studied by sociology. I say this in deliberate contrast to the widely held view that most of our problems will be solved by technology. I have discussed this point elsewhere (Babbie 1988), so I will give only a few illustrations here:

◆ Overpopulation is an enormous problem for our planet. Although the invention of a variety of effective contraceptive methods is to be welcomed, these methods have not solved the problem. The solution lies in the realm of values and norms.

◆ Agricultural advances have not ended the tragedy of world hunger, which kills 13 to 18 million people every year. In fact, we now produce more than enough food to feed everyone on the planet, and local famines account for only about 10 percent of the annual deaths due to hunger. The remain-

ing 90 percent are a consequence of chronic, persistent
hunger, the solution to which lies in socioeconomic structures and in the social psychology that keeps us from doing
what is necessary.

◆ Finally, we have spent much of our history in the search for
weapons that will make war unacceptable or unthinkable,
but we have merely expanded our ability to kill on a grand
scale. The solution to war, again, lies in the domain
addressed by sociology.

I do not mean to suggest that sociology currently has all the
answers to all the problems of the world. I do suggest, however, that sociology is the place to look for answers to a great
many of those problems. In this sense I suggest that sociology
is an idea whose time has come.

References

Babbie, E. 1986. *Observing Ourselves*. Belmont, Calif.: Wadsworth.

Babbie, E. 1988. *The Sociological Spirit: Critical Essays in a Critical Science*. Belmont, Calif.: Wadsworth.

Babbie, E. 1989. *The Practice of Social Research*. Belmont, Calif.: Wadsworth.

Gall, J. 1975. *Systemantics: How Systems Work and Especially How They Fail*. New York: Quadrangle.

Kuhn, M., and T. McPartland. 1954. "An Empirical Investigation of Self-Attitudes." *American Sociological Review* 19:68–76.

Maturana, H. 1980. *Autopoiesis and Cognition: The Realization of Living*. Dordricht, The Netherlands: D. Reidel.

Stouffer, S., et al. 1949, 1950. *The American Soldier*. Princeton: Princeton University Press.

Earl Babbie is vice president for Research and Planning at
Chapman University. Prior to his current position, he served as chair
of the Social Science Division and still teaches research methods. He
is the author of several textbooks; best known of these is *The Practice
of Social Research*. He is a member of the ASA Teaching Committee
and the Membership Committee. Address correspondence to Earl
Babbie, Chapman University, Orange, CA 92666. Compuserve:
76424,156.

COMMENT ON "THE ESSENTIAL WISDOM OF SOCIOLOGY"

James A. Davis

Harvard University

I assume I was invited to comment because the editor believed I would enliven things with a savage attack. Since I have long harangued the profession on the desirability of building empirical research into introductory sociology, the assumption seemed reasonable.

Try as I may, however, I cannot work up a really impressive level of vitriol because I rather liked the paper. I'm for almost any approach that gets us away from the namby-pamby Reader's Digest Encyclopedia of Mush ladled out in our intro texts. I am also delighted to see serious discussion of content as well as pedagogical methods in *Teaching Sociology*. And any statement of heavy sociological truths that includes a risque joke cannot be all bad.

As a socialized sociologist, however, I must conform to my role—hired assassin—and I do not find this impossible. Two reservations come to mind. Economists would call them "comparative advantage" and "demand."

While Professor Babbie's Ten Commandments summarize ideas and assumptions that permeate sociology, they have little to do with actual sociology as it is practiced. (In fact, he tells us sociologists have never heard of #3.) Assuming sociology to be an academic field, its intellectual life is in the journals and professional meetings. The most casual inspection of journals

James A. Davis, "Comment on "The Essential Wisdom of Sociology,'"
Teaching Sociology, October 1990, pp. 531–532.

and meeting programs shows they are mostly about empirical research, occasionally about substantive theories, and almost never about metaconcepts. It thus seems odd that instruction should center on a collection of less than rigorous working assumptions and not on their thousands of concrete payoffs. Economists and psychologists have working assumptions which are quite different. This doesn't mean they or we are wiser. What is important is not the truth of the assumptions (it can't be tested directly), but whether they pay off in findings. Thus, to me Babbie's propositions are all foreplay and no sex.

As we mourn the premeditated murder of sociology in two medium-quality universities, it may be prudent to consider our competitive position in the jungle of modern academia. Zoo may be a better metaphor. Unless our plumage is more colorful or our cubs more cuddly than those in the cages of neighboring disciplines, we may find our keepers in the deaneries reluctant to dish out the birdseed or horse meat or whatever. I am not persuaded that "wisdom" is our brightest plumage or cutest cub.

In adjacent cages, philosophy, history, American studies, and the other behavioral social sciences are also preening, begging, and assuming. They have plenty of wisdom and I can't see it is inferior to ours—if only because wisdom in the form of grand verbal assumptions is almost impossible to vet. I'm not knocking our wisdom; it's pretty good as wisdom goes, but it doesn't meet the principle of comparative advantage: "Do not do what you are good at, do what you are better at than the competition."

Where is sociology's *comparative* advantage? To me it's obvious. Unlike our academic competitors, we know things about contemporary reality. We get out there, we talk to real people, we draw our conclusions from data, not from novels about sensitive young intellectuals, artificial indoor experiments, inscrutable thoughts of recherché European pundits, the memoirs of retired diplomats, or ideology (well, we used to eschew ideology, and I see signs we might be returning to our senses). In short, we see Price's (1969, p. iii) paradox again. Let me present it to the readers of *Teaching Sociology* a second time:

> **The feature of contemporary sociology that is perhaps its greatest strength—its solid factual base—is under-represented in introductory sociology textbooks and anthologies, whereas the features of relative weakness—its concepts, propositions, and theory—are over-represented (Davis 1978, p. 235).**

In sum, the principle of comparative advantage leads me to hesitate.

I am less sure of a second problem, but sure enough to sketch it. I'm doubtful these ideas will seem all that novel to today's students. When I took intro in 1947, I came to it brain-washed by the lily white, crackpot anti-Roosevelt individual-ism of small town, Midwestern America. Sociology was an eye opener for me! But I fear we have been so successful our students come as pre-shrunk Babbie-ists. They already "know" about cultural relativism, they already "know" about the sys-tem and how awful it is. They "know" society presses down (on other people) from the outside, they believe the polls. Indeed, my students seem to hold a rather complex ideological system that maintains that (1) everyone in the world, except white, middle-class Americans, is shaped by *her* culture and nothing they do should be judged harshly, while (2) white, middle-class Americans (including roommates' parents, but not their own) should be held personally accountable for all that is wrong with the world. This folk wisdom—a mélange of youthful idealism, pop social science, and diluted sixties left-ism—is not exactly Babbie-ism, but it is so close I doubt con-temporary students will gasp when encountering the ten commandments. If our zingers don't zing, look out for the zoo-keeper. Selling to a saturated market is not the best strategy, even if our product were superior.

In sum, I believe the Babbie Ten are a legitimate part of sociology and might well make a nice framework in which to arrange concrete, definite information. But I am less than enthusiastic about considering "wisdom" as the highlight of our teaching.

I am not saying the ten commandments of wisdom are wrong, just that putting them forward as our salvation is, uh, unwise.

Verbum sap.

References

Davis, J. A. 1978. "Using Computers to Teach Social Facts." *Teaching Sociology* 5:235–257.

Price, J. L. 1969. *Social Facts.* New York: Macmillan.

James A. Davis is professor of sociology at Harvard University and principal investigator, NORC General Social Survey. In 1989 he received the ASA Distinguished Contributions to Teaching Award for his fanatic advocacy of data analysis in introductory undergraduate courses. Address correspondence to James A. Davis, Department of Sociology, Harvard University, Cambridge, MA 02138. Bitnet: JAD@HARVUNXW.

PRIORITIZING SOCIOLOGICAL PERSPECTIVE OVER CONCEPTS

Michael R. Leming

Saint Olaf College

One of our department's annual spring rituals is to interview our "best" senior majors and decide which (if any) will graduate with departmental distinction. Prior to the interview, students are required to submit a portfolio of term papers written in their sociology courses along with a short autobiographical statement tracing the development of their sociological development. This spring three seniors were seeking the departmental honor. In the course of our interviews we asked each of these students to describe the sociological perspective. One might think that any student with a major in sociology would be able to answer this basic question. However (to our great dismay), such was not the case. All three students described sociology in terms more suited to the discipline of psychology.

Certainly, our students know the proper words to describe the discipline—they had to learn these words in order to pass introductory sociology and to defend their choice of this major to their parents. But when asked to characterize the sociological perspective, they quickly became tongue-tied and confused. As our faculty encountered the "best" products of our teaching, an existential angst fell over the department.

My first response to this problem was to "blame the victim." These students simply missed the mark. They did not internalize the eternal varieties of the discipline they were

Michael R. Leming, "Prioritizing Sociological Perspective over Concepts," *Teaching Sociology,* October 1990, pp. 533–535.

taught. How could they be so dull? My second response was to blame another victim—the department (since I had not taught any of these students introductory sociology). Where had we, as a department of "exceptional" teachers, gone wrong? Did our present students reflect the quality of our sociological instruction?

"No," I said. "I'm too much of a sociologist to accept these simple answers." Neither the students nor faculty are responsible for this state of affairs. The fault lies in the individualism of American culture. If our students arrive on campus having internalized an intellectual perspective predicated upon a sort of "pop" psychological reductionism, why wouldn't they naturally see sociology as a psychology of culture, social structure, and group behavior?

During this past year I encountered evidence that supports this explanation. First, there is the matter of the sociologist's view of psychology. Members of the college's department of psychology and their students complained to me (as chair of the department of sociology) that our faculty are perceived on campus as "psych bashers" (a characterization that is probably supported by considerable evidence). In our heavily "psychologized" world, serious sociology is bound to be perceived as psych bashing. Sociologists radically challenge the taken-for-granted psychological perspective when they claim that culture, group behavior, and social structure cannot be reduced to individual actions. Sociologists consistently teach that psychology is of limited value if one is concerned with social interaction. Second, consider this fact. Most professors of sociology think that Peter Berger's *Invitation to Sociology* and C. Wright Mills's *Sociological Imagination* are great books. And yet, most undergraduate students dislike them. Why is that? If we discount the fact that the vocabulary encountered in each of these books is beyond many contemporary undergraduate students, the primary reason students do not like these classic books is that they do not want to think about their world in sociological terms. Students find it more comfortable to view sociology as the psychology of group behavior.

When challenged by the sociological perspective, our students have three options:

1. Leave sociology for the safer confines of psychology and complain that the sociologists are "psych bashers."
2. Stay in sociology but reformulate the discipline (its concepts, perspectives, and theories) in a manner compatible with psychology.
3. Abandon the psychological orientation and "convert" to a sociological perspective.

When I recall my own sociological background and the educational biographies of most of my professional colleagues, sociology only began to "make sense" in graduate school. One reason for this might be that as one acquires an undergraduate sociological education, one learns the vocabulary of the discipline without internalizing the perspective. Sociological concepts are the tools of the discipline, but there is no guarantee that they will be used in the manner intended. With little effort these concepts can be reformulated in psychological terms.

For the past 20 years it has been my contention that one cannot understand the sociological perspective apart from the language of the discipline. But I now recognize that disciplinary vocabulary is a necessary, *but not sufficient,* cause for understanding the sociological perspective. Unfortunately, our texts are written as if our disciplinary language is *all* that is needed to understand our discipline. Introductory textbooks stress vocabulary and "key concepts." Instructor's manuals emphasize disciplinary terms in the construction of objective and essay test items.

Here is where I come to Earl Babbie's article, "The Essential Wisdom of Sociology." If students' first encounter with sociology emphasized the unique perspective and assumptions of the discipline, we might be able to convince them "that sociology is an idea whose time has come." Although I think that sociology has contributed more than "Ten wisdom essentials" (and I think Babbie needs to provide more elaboration of his "essentials"), I believe Babbie is on the right track in the transformation of the teaching of sociology. His introduction makes the point that I am arguing—*perspective* should take priority over *concepts*. According to Babbie:

> **It is important for students to grasp some fundamental sociological concepts: role, interaction, structure, conflict, socialization, etc. . . . [but it is] even more important to ground students in some basic principles or metaconcepts that distinguish the sociological view of things and make sense of the concepts we weave together in the sociological construction of society.**

It is for this reason that I spend the first two weeks of my introductory sociology course considering Peter Berger's explanation of the social construction of social reality. In "Religion and World Construction" (the first chapter of *The Sacred Canopy*), Berger makes the following points, all of which are similar to Babbie's metaconcepts (Babbie's points are numbered in parentheses):

1. Humans are without instincts and therefore must make a "world," or normative order for themselves. This order (what Berger calls a "nomos") consists of the group's culture and social structure.
2. The nomos is a human construction, it is characterized by a built-in instability. Humans do not have a given relationship to the nomos, and therefore must continue to establish a relationship with it through the process of socialization and resocialization (points 4 and 6).
3. Society perpetuates the social order through the socialization process *(Autopoiesis,* point 3). The individual not only learns the objectivated meanings of society but is shaped by them (point 9).
4. The social order is a human product that acts back on its producer (point 1). In creating a social order, and locating their own biographies within that order, humans not only provide order for society but also for themselves (point 5).
5. Once produced, culture or social order (nomos) cannot be wished away because it stands outside of the subjectivity of the individuals who produced it *(Sui Generis,* points 1 and 7).

6. The socially constructed world is above all an ordering of experience. A meaningful order, or nomos, is imposed upon the discrete experiences and meanings of individuals (point 9).

7. If the social order is objectified and is taken for granted by society's members, it can impose itself upon reluctant individuals and become the most effective agent of social control.

8. The social order has a tendency to expand into wider areas of meaning—even the future attains meaning by virtue of the "nomos" being projected into it. The Cosmos is a social order that is taken for granted and is projected into the future. Religion is the human enterprise by which the sacred cosmos (the transcendent projected nomos) is established.

9. The most important function of the cosmos may be understood as society's shield against terror. It is in the marginal situations of life—death, war, and natural disasters—that the cosmos makes stable the group's way of life. This is possible because the cosmos is not perceived as being socially constructed but rather as the "fundamental meanings of the universe."

While Berger's essentials do not deal with the science of sociology (Babbie's points 2 and 8), I believe that they do provide students with an understanding of the uniqueness of the sociological perspective. It is my opinion that most (if not all) introductory sociology textbooks subordinate the perspective of the discipline to sociological vocabulary. Babbie's "essentials" are a good first step in correcting this problem.

References

Berger, P. L. 1963. *Invitation to Sociology*. New York: Anchor.

Berger, P. L. 1967. *The Sacred Canopy: Elements of a Sociological Theory of Religion*. New York: Doubleday.

Mills, C. W. 1959. *The Sociological Imagination*. New York: Oxford University Press.

Michael R. Leming is professor and chair of the Department of Sociology at St. Olaf College. He is coauthor (with George E. Dickinson) of *Understanding Dying, Death, and Bereavement* (2nd ed.), coauthor (with Raymond DeVries and Brenda Furnish) of *The Sociological Perspective: A Value-Committed Introduction,* and co-author (with George E. Dickinson) of *Understanding Families: Diversity, Continuity, and Change.* He is the founder and former director of the St. Olaf College Social Research Center, a member of the board of directors of the Minnesota Coalition on Terminal Care, and a steering committee member of the Northfield AIDS Response. In addition, he serves as a hospice educator, volunteer, and grief counselor, and is the secretary-treasurer of the Christian Sociological Society. Address correspondence to Michael R. Leming, Department of Sociology, St. Olaf College, Northfield, MN 55057. Bitnet: LEMING@ACCSTOLAFEDU.

RESPONSE TO BABBIE

Laura Kramer

Montclair State College

Reading Babbie's reflections is a validating experience for one who routinely teaches sociology to nonmajors and those taking it for the wrong reasons (e.g., it's the only course open at the right time that satisfies a requirement and is assumed to be a "gut"). It pushes me to articulate and review those ideas central in my attempts to convert students to my belief in sociology's essential wisdom. In a more reactive mode, these ideas are crucial when faced with the challenge of a student's implicit or explicit "So what?" response to course material.

Some of the "basic principles" are old friends, and others on the list have provoked me to make my own variations explicit. First, some comments on Babbie's principles.

1. *Society has a* sui generis *existence*. This is one of the most difficult points to teach. Indeed, while the Supreme Court *case* on racism and capital punishment exemplifies the existence of society, the Court's *ruling* exemplifies the blindness of most Americans to society's existence *sui generis*. Even after a student sees the light with a personally effective example, regression to the individualistic level of explanation is common (sociologists are not, ourselves, immune to this!); I have found collecting and updating examples of this principle useful. It is even more effective to have students look for their

Laura Kramer, "Response to Babbie," *Teaching Sociology*, October 1990, pp. 536–537.

own examples in newspapers, movies, the campus, or the workplace. Some of the most successful examples deal with personal troubles/public issues in their own lives.

2. *Studying society scientifically.* I appreciate the use of *The American Soldier;* I assume that each of us has a favorite set of findings to illustrate the inadequacy of common sense and the power of systematic inquiry. Like many, I use group differences in suicide rates at the beginning of my introductory course. However, I prefer not to misrepresent findings to convince students of the inadequacies of "common sense." Instead, I pose questions (e.g., Do you think there are differences in suicide rates among religious groups? Which religions do you think have higher or lower rates?), and ask students to articulate the basis of their answers. Consummate style is needed to bring off misrepresenting findings without embarrassing students who have pontificated about false findings, and might be reluctant to participate in subsequent discussions.

A crucial part of teaching this principle is encouraging a critical perspective toward empirical work. After discussing official suicide rates, I raise some of the basic problems with such rates (e.g., if a family's religion forbids burial in sacred ground to suicides, the family will try harder to have the death recorded differently). Of course, (being a positivist) I then try to show how we minimize sources of error in order to better pursue a scientific study of society and social phenomena. Without this final piece, students may feel authorized to toss out the results of empirical research without reason.

5 and 8. The *inseparability* of individual and society, and the *deterministic* approach of explanatory sociology are more difficult to teach than might first appear. Students are quick to perceive the power of social context when they are focusing on human action of which they disapprove, but find it harder to see any role for social systems in shaping desirable outcomes. Examples of negative influences are easier to find, and possibly more fun to teach with, but should be balanced with positive examples. Otherwise, many students continue to assume

that a "good society" is simply one that doesn't interfere with the emergence of the goodness in people's selves, having no active formative role in it.

9. *Paradigms* reveal and distort. This principle, well- and often-illustrated, should help to reduce students' propensity to toss out sociology because it cannot predict perfectly and cure totally. I believe there is significant resistance to sociology from students who are "shooting the messenger": we present bad news, and don't often report success (either in designing positive social change or accurately predicting the future). Rather than being defensive about a lack of one grand and all-encompassing theory, we can present our multiple paradigms positively: when one fails, another yields insights.

10. *Sociology's time has come.* I think sociology's time has been here for quite a while—at least since the social world that led to its development in the nineteenth century. I agree that it is "the best way" to tackle the present and improve humanity's future chances.

To make this idea compelling, I need to know why we should expect the value of sociology to be recognized more today than during the preceding decades of major social problems. In fact, this idea strikes me as being of a different sort than the other principles—more of an evaluation of the sum of the first nine principles than an additional tool for understanding human life.

To Babbie's list I would add one of my own—*Social life is dynamic*. While asserting the dynamic nature of social life does not conflict with, indeed is implicit in, several of his, I believe it needs to be stated as a separate principle. It fits neatly paired with, or counterbalancing, the principle of institutional conservatism. I think many of us still have to work on integrating a dynamic view, beyond simply asserting it from time to time as we teach.

This principle is easier to teach now than it has been for some years; well-publicized examples of changes in South Africa, Eastern Europe, and rapidly growing concern with the environment in the United States make this perspective more

believable than previously for students raised during the
Reagan era. Students need to see the social processes leading
to change, however. Media discussions tend to present the
changes as inevitable trends toward goodness and away from
evil, as proof of the rightness of our system. Instead, we need
to emphasize the importance of structural and cultural forces in
effecting change.

Too often, descriptions of the processes of social change
are segregated in a chapter on that topic; all other topics are
taught with the presumption of stasis. For example, organizations are classically described as having essentially stable
power distributions, even if they often depart from formal
descriptions. Coalitions, and even their own fluidity, are inte-
gral in a dynamic treatment of organizational structure and
process, rather than mentioned as an afterthought. Perrow's
suggestion (1979) that the gap between formal rules and current practice is normal, rather than pathological departure from
the ideal-type, also exemplifies the use of a dynamic model.

Many of us continue to teach using models that assume stability, and only tack on "new ideas" as a final topic. In presenting socialization, even when the oversocialized conception of
"man" (Wrong 1961) is discussed, the fundamental view of the
individual is essentially reactive. As we teach students how
people are socialized, we prepare them to expect social change
to be impossible.

In other words, if we are to teach the principle that social
life is dynamic, we must rethink some of our favorite paradigms, examples, and lessons. This requires re-visioning, looking differently at the world that we have long believed we
understood. If we aren't really convinced, we won't convince
our students with a few "canned" lectures. The catechismic
nature of our presentation of this principle will be apparent
when we revert to our static assumptions in discussion.
Pragmatically, teaching this principle involves a lot of work in
the development of course materials. However, as I believe we
see in feminist sociology, the excitement that develops from
changing paradigms invigorates teaching as well as stimulates

and improves the validity of other intellectual activity (see, for example, Daly and Chesney-Lind [1988] on changing paradigms in criminology and Thorne [1982] on the family).

Our students get a well-grounded appreciation for the forces contributing to social order and social cohesion. Without abandoning those lessons, we should strengthen their education in the dynamic nature of social life.

References

Daly, K., and M. Chesney-Lind. 1988. "Feminism and Criminology." *Justice Quarterly* 5:497–538.

Perrow, C. 1979. *Complex Organizations: A Critical Essay.* Glenview: Scott Foresman.

Thorne, B. 1982. "Feminist Rethinking of the Family: An Overview." Pp. 1–24 in *Rethinking the Family,* ed. B. Thorne with M. Yalom. New York: Longman.

Wrong, D. 1961. "The Oversocialized Conception of Man in Modern Sociology." *American Sociological Review* 26:183–193.

Laura Kramer is associate professor and chair of the Department of Sociology at Montclair State College. Interests include gender, technology, and work. She authored *The Sociology of Gender: A Text-Reader.* Address correspondence to Laura Kramer, Department of Sociology, Montclair State College, Upper Montclair, NJ 07043.

MAKING SOCIETY (AND, INCREASINGLY, THE WORLD) VISIBLE

John J. Macionis

Kenyon College

One evening last spring, a 28-year-old woman was jogging through New York's Central Park when she was attacked by six young boys. The band raped and brutally beat the terrified woman, then left her for dead. Discovered three hours later, she had lost half of her blood and was clinging to life in a coma. Miraculously, she survived and has made a remarkable and courageous recovery. She is now back at her job on Wall Street.

Perhaps the most intriguing element of the "wilding" case was the inability of the six young men to account for their actions. ("Wilding," as you will recall, was the way one of the boys described the group's attacks on strangers: "We were just wilding . . .") More important for the present discussion, what attracted national attention was that the "wilding" episode provoked a spirited exchange in the press as to the merits of sociology (Macionis 1991). In the wake of the attack, a number of social scientists offered various insights about the episode. Columnist George Will (1989) dismissed as "psycho-socio babble" explanations of the malicious attack by what he thought were well-meaning but misguided social scientists. Sociologists, Will charged, only serve to confuse the issue through a "dispersal of responsibility in a cloud of socioeconomic factors." Some other Americans would probably have

John J. Macionis, "Making Society (And, Increasingly, the World) Visible," *Teaching Sociology*, October 1990, pp. 538–539.

used more direct language, blasting sociologists and other analysts as "bleeding hearts" who strip personal responsibility from what, in this instance, seemed to be a clear-cut case of brutality.

I think that, properly understood, sociology need make no apology. On the contrary, the wilding incident and its aftermath help to clarify the power and the value of a sociological perspective. In fact, this event provides a useful opportunity to reflect on what the "essential wisdom of sociology" might be.

To most Americans, and certainly to George Will, the explanation of the "wilding" attack was simple enough: bad people. No doubt a passionate sense of injustice is best served by this kind of simplicity. That is, when things go right, and especially when they go wrong, we understandably want to place responsibility. In most modern, industrial societies such as the United States, this typically means pointing a finger at particular individuals. For me, the power of sociology lies in challenging this kind of thinking.

Of course, sociology does tend to take a deterministic view of human behavior; given our cultural nature, how could we do otherwise? We should therefore turn Babbie's observation about sociology around and remark on modern Americans' highly personal view of human behavior. Put another way, over the past centuries, "society" seems to have become less and less visible. The essential task of sociology, then, is to make society visible so that we can comprehend how individuals think, feel, and act within a context of social forces.

Consider how the "wilding" episode takes on greater meaning when examined sociologically. Would anyone imagine that girls are as likely as boys to commit this kind of crime? [In the Unites States, men commit nine times as much violent crime as women do.] What about middle-aged people? [Americans between fifteen and twenty-four years of age are 15 percent of the population, yet account for almost half of violent crime.] More broadly, crimes of this kind are also remarkably *American:* there are more deaths annually from stray gunshots in New York than there are intentional murders in most European cities.

This is the power of thinking sociologically. Is sociology's claim that crime is the result of a complex array of factors the unrealistic claim of a "bleeding heart," as Mr. Will would maintain? Hardly. On the contrary, a sociological analysis seems far more realistic than his contention that some people are simply "evil." We need to keep in mind, of course, that the analytical truth of sociology and moral or legal truth need not be the same. [I am reminded of one of my own students who, several years back, was called before the dean for misbehaving, and offered a Durkheimian account of the *need* for deviance in any healthy social setting.] But there can be little doubt about the need for us, as contemporary Americans, to incorporate a little sociological awareness in our thinking.

Our task in introducing sociology to students, then, is to acknowledge that people make choices, to bear in mind that we often hold people personally responsible for what they do, but above all to reveal through research that individual activity is responsive to the complicated arena of human society.

I think, too, that we are beginning to see that placing the individual within the context of American society realizes only part of the promise of sociology. We must also recognize that we live within an increasingly interdependent world order. Just as the operation of American society guides the experiences of individuals, so the global systems of culture, politics, and economics are linked to what happens in the United States. An added task as we approach the next century, I think, should be to help extend the "reach" of sociology to this world context. Currently, there are more than 350,000 foreign students in the United States, while only 50,000 of our own students (out of 12 million) study abroad each year (C.I.E.E. 1988). Why does this matter? There are some very practical reasons. Consider that one-third of American corporate profits are earned abroad, four-fifths of new American jobs are created as a result of foreign trade, Americans have over $300 billion invested in other

countries, and foreigners have invested $1.5 trillion in the
United States (C.I.E.E. 1988). Sociology could make an especially significant contribution to the global sophistication of
Americans if the 750,000 students who enroll in an introductory course each year were provided with greater understanding
of how American society is linked to the world as a whole.

I am largely in agreement with Babbie's intention in
putting together his ten points (actually six about society and
four about sociology). We will, as he concedes at the outset,
argue constructively about precisely what constitutes the
"essential" wisdom; Babbie's listing seems likely to attract
more favor from structural-functionalists than from interac-
tionalists or conflict sociologists. Yet I hope we will all push
beyond some of the particulars toward agreement on the
broader goal of making visible the structures and processes
that inevitably shape human activity—and to do so with increasing attention to the various ways in which American society is bound up with the entire world.

References

Council on International Educational Exchange. 1988. *Educating for
    Global Competence: The Report of the Advisory Council for
    International Educational Exchange.* New York: C.I.E.E.
Macionis, J. J. *Sociology.* 3rd ed. 1991. Englewood Cliffs, NJ:
    Prentice-Hall.
Will, G. F. 1989. "No Psycho-Socio Babble Lessens the Fact that Evil
    was the Crux of Central Park Rape." *The Philadelphia Inquirer,*
    May 1.

John J. Macionis is associate professor of Sociology at Kenyon
College. He has written *Sociology,* coedited *Seeing Ourselves: Classic,
Contemporary, and Cross-Cultural Readings in Sociology,* and coau-
thored *The Sociology of Cities.* Address correspondence to John J.
Macionis, Department of Anthropology and Sociology, Kenyon
College, Gambier, OH 43022.

RESPONSE TO THE COMMENTS ON "ESSENTIAL WISDOM"

Earl R. Babbie

Chapman University

My purpose in preparing "The Essential Wisdom of Sociology" was to stimulate a conversation, so I was delighted by the comments of Davis, Leming, Macionis, and Kramer. With no intention of rebutting anything they said, I appreciate the opportunity to have a few last words.

When Jim Davis begins turning phrases like "namby-pamby Reader's Digest Encyclopedia of Mush" and "deaneries reluctant to dish out the birdseed," I'm truly grateful that he "rather liked the paper" and couldn't work up a lot of vitriol about it. His comments about the empirical aspect of sociology are so important, however, that I wish I'd expanded my scope beyond the realm of concepts and metaconcepts to include that. God, I'm embarrassed by even the appearance of ignoring research.

Mike Leming is right in saying there are more than ten items of essential wisdom in sociology. Counting Jim Davis's point about our empirical basis, there are at least eleven. Mike's review of Peter Berger's work adds even more, though Mike is very kind in relating most of them to my original ten.

I liked Laura Kramer's suggestion of adding "Social life is dynamic." That makes an even dozen, Mike. Satisfied?

John Macionis makes a powerful case for the importance of sociological points of view in confronting the problems of the day. The contrast he draws to the George-Willian-view nicely summarizes the task we face in communicating to the general public.

John also suggests that my points may have a structural-functional bias. This isn't true for me, but John's not the only one to make that comment, so readers of the original piece should be warned of that possibility. Admittedly, I first learned sociology as an undergraduate from Talcott Parsons (who had a certain fondness for functionalism), and I originally thought he invented sociology. Today, my brain more commonly haunts a territory somewhere to the left of ethnomethodology, though evidently my flashing fingers favor functionalism (and also alliteration).

GLOSSARY

achieved status Social position gained through your actions, such as college graduation or ax murder. Contrast with **ascribed status**.

aggregation As distinguished from **group**, this refers to a gathering of people in the same location, such as the people gathered to watch you let the air out of your instructor's tires. (Money will probably weaken their memories.)

alignment That quality marking a group of individuals who share a common purpose, with each doing whatever is appropriate to achieve it.

ascribed status Social position established at birth, such as sex or race. Contrast with **achieved status**.

attribute Characteristic of a person or thing. A set of logically related attributes make up a **variable**.

belief Reified view of what's true, usually shared as an agreement among members of a **group**.

bureaucracy (1) Form of social organization characterized by specific procedures and status relationships. (2) Government in the hands of a clique of cabinetmakers.

concepts The mental images we use to bring order to the mass of specific experiences we have.

conflict paradigm Theoretical point of view that focuses on social disagreements and the struggle among individuals and groups to impose their points of view on others. Stresses the role of power in social affairs.

cultural relativity Recognition that the agreements different groups share are only different, with none being better or worse than another. Contrast with **ethnocentrism**.

culture (1) Shared symbols, beliefs, values, norms, and other agreements of a group such as a society. (2) What you were sent to couth school to get. (3) What you wash your ears to avoid.

ethnocentrism Feeling that the agreements of your group are real, true, and better than the agreements of some other group.

experiment Research method involving the controlled manipulation of variables, often in a laboratory situation. Evaluation research involves "natural experiments."

extended family See **family**.

family Social institution dealing with the group need for replacement of members through reproduction.

> **extended family** Unit consisting of parents, grandparents, and children.

> **group marriage** Several husbands and wives live together as mates to each other.

> **nuclear family** Unit consisting of mother, father, and children.

> **polyandrous marriage** One wife has several husbands.

> **polygynous marriage** One husband has several wives.

functionalist paradigm See **social systems paradigm**.

generalized other Term used by George Herbert Mead in illustrating how we learn about overall community agreements. Having looked at things through the points of view of many different people, we develop a sense of how people in general—the generalized other—see things.

group A plurality or collectivity of people who share common interests, interact with one another, have a sense of identity with one another, and have some degree of structure.

group marriage See **family**.

I The acting and reacting aspect of the self. Contrast with **me**.

in-group Group you feel you belong to but exclude other people from. See also **out-group**.

institution Set of established and persistent agreements governing a broad aspect of social life. Sociologists use this term to refer to such things as the family, religion, education, politics, and economics. The primary function of an institution is to shape individual experiences in such a way as to support the survival of the group.

institutional prejudice Situations in which prejudice and discrimination exist within a society or a large portion of society, yet no individuals can be identified as the culprits.

interaction The process in which one person directs a communication and evokes a response from the other that is conditioned by the initial communication. Realize in this that "communication" includes physical actions as well as words. Typically, it goes back and forth, back and forth, like a windshield wiper. Also, any number can play.

interactionist paradigm The sociological perspective that focuses on social life as the process of give and take, in which individuals come to grips with each other, forming common definitions of the situations they find themselves in.

internalize To take inside ourselves and make our own.

labeling theory The sociological view that people often take on the statuses assigned them by the opinions of those around them.

logical/empirical To be accepted, scientifically valid assertions must make sense and correspond to the facts.

looking-glass self (1) Term coined by Charles Horton Cooley to refer to the way you see yourself as a reflection of the way other people seem to see you. (2) That friend you talk to in the bathroom mirror when you think no one is watching.

macrosociology The sociology of large-scale structures and processes, such as whole societies.

me The aspect of the self that reflects the views of others. Contrast with **I**.

microsociology The sociology of small-group interactions and similar small-scale phenomena.

minority group Subset of society, such as women or blacks, that receives a disproportionately small share of society's rewards.

negative feedback loop A system whose elements tend to hold each other in check, such as price and demand in economic systems.

norm (1) Agreement on the behavior expected of people; shaking hands when meeting someone is a norm. (2) Residential building for college students with colds.

nuclear family See **family**.

organization A set of agreements or rules that form a context within which people interact to achieve certain objectives.

out-group (1) Those people who are excluded from a particular **in-group** or from the rest of society. (2) Unsuccessful batters in a baseball game.

paradigm Model or perspective.

pluralism A social pattern that provides for the coexistence of different subcultures within the same society.

polyandrous marriage See **family**.

polygynous marriage See **family**.

positive feedback loop A system whose elements reinforce each other in a runaway spiral, such as wages and prices in economic systems.

primary group Group characterized by **primary relationships**. Now guess what a **secondary group** is.

primary relationships (1) Close, intimate relationships, such as those among friends; characterized by a "we-feeling." (2) Political hanky-panky.

recursive. The equality whereby a process turns back on itself; for example, if we were to learn of a beautiful, deserted beach, that knowledge would lead people to go there—and it wouldn't be deserted any more.

reference group Those people you compare yourself with or model yourself after.

reification Treating things as real when they are not.

role The behavior expectations associated with a social **status**. The things you are expected to do because of the status you occupy; for example, students are expected to study.

sanctions Rewards and punishments.

secondary group Group characterized by **secondary relationships**, such as an association.

secondary relationships Casual, impersonal relationships (contrast with **primary relationships**); usually aimed at accomplishing a specific purpose.

self Who you are, or at least who you think you are. This is sometimes used to mean *identity*. See also **I** and **me**.

social category As distinct from **group**, this refers to people who merely share a particular characteristic in common, such as left-handed people.

social control Enforcement of a group's agreements, accomplished by a variety of mechanisms and agencies.

socialization The process of learning the rules of society.

social structure Established network of relationships connecting different statuses in a group, including the norms for interactions among different statuses.

social systems paradigm The sociological perspective that focuses on the structure of social life, especially seeing society as an integrated system in which each individual has a part to play.

society A voluntary association of individuals for common ends.

status A position or location a person can occupy within society or a smaller social grouping.

stratification The structuring of inequalities in society; for example, the ranking of people as upper class, middle class, and lower class.

survey research The use of questionnaires to collect information.

theory (1) Logical model specifying relationships among variables as an explanation for the way things are. (2) A wish, as in, "he had a *theory* that the garbage truck would stop for a Mercedes."

truth Sincerity in action, character, and utterance.

values A matter of preference, representing views about what's better than what.

variable Logical grouping of **attributes**. "Sex" is a variable made up of the attributes "female" and "male."

SUGGESTED READINGS

Berger, Peter L. 1963. *Invitation to Sociology: A Humanistic Perspective*. New York: Doubleday.

Berger, Peter L., and Luckman, Thomas. 1966. *The Social Construction of Reality*. New York: Doubleday.

Bottomore, T. B., ed. & trans. 1976 [1843]. *Karl Marx: Selected Writings in Sociology & Social Philosophy*. New York: McGraw-Hill.

Durkheim, Emile. 1951 [1897]. *Suicide*. Glencoe, Il.: The Free Press.

Gerth, Hans, and Mills, C. Wright, eds. 1946 [1925]. *From Max Weber: Essays in Sociology*. New York: Oxford University Press.

Goldsmid, Charles A., and Wilson, Everett K. 1980. *Passing on Sociology*. Belmont, Calif.: Wadsworth.

Horowitz, Irving Louis, ed. 1969. *Sociological Self-Images*. Beverly Hills, Calif.: Sage.

McClung, Alfred Lee. 1973. *Toward Humanist Sociology*. Englewood Cliff, N.J.: Prentice-Hall.

Mead, George Herbert. 1934. *Mind, Self, and Society*. Chicago: University of Chicago Press.

Nisbet, Robert. 1966. *The Sociological Tradition*. New York: Basic Books.

Parsons, Talcott. 1954. *Essays in Sociological Theory*. New York: The Free Press.

Weber, Max. 1958 [1905]. *The Protestant Ethic and the Spirit of Capitalism,* trans. Talcott Parsons, New York: Charles Scribner's.

INDEX